Mariëtte Boon (1988), MD, PhD is internal medicine specialist in training. Her research, performed at the Leiden University Medical Center, Leiden, focuses on fat metabolism. She has been awarded numerous prestigious national and international prizes for her research. Her favourite sport is tennis and her favourite food chocolate.

Professor Liesbeth van Rossum, PhD is internist-endocrinologist at the Erasmus University Medical Center, Rotterdam. She is co-founder of Obesity Center CGG, and has an internationally leading position in the field of obesity and biological stress research. Van Rossum is a frequently invited speaker, including a TEDx talk, and received over 20 international awards for her scientific research. She's also an athletics fanatic.

fat
The *Secret* Organ

MARIËTTE BOON &
LIESBETH VAN ROSSUM

Translated from the Dutch
by Colleen Higgins

Quercus

Originally published in 2019 in The Netherlands
by Ambo Anthos Uitgevers under the title *Vet Belangrijk*.

Trade paperback edition published in Great Britain in 2020 by
Quercus Editions Ltd

This paperback edition published in 2021 by

Quercus Editions Ltd
Carmelite House
50 Victoria Embankment
London EC4Y 0DZ

An Hachette UK company

PB ISBN 978 1 52940 091 5
Ebook ISBN 978 1 52940 092 2

10 9 8 7 6 5 4 3 2 1

Typeset by Jouve (UK), Milton Keynes

Printed and bound in Great Britain by Clays Ltd, Elcograf S.p.A.

MIX
Paper from
responsible sources
FSC® C104740

Papers used by Quercus Editions Ltd are from well-managed forests and other responsible sources.

To our parents

TABLE OF CONTENTS

FOREWORD

When you hear 'body fat', you probably think mainly about that little bulge that starts to peek out over the top of your jeans when you've eaten too much over the holidays. Or about that fat on your backside that shows dimpling when viewed in the wrong light (in most women, anyway). Many people have a love-hate relationship with their body fat, and in most cases, hate usually has the upper hand.

This attitude is clearly reinforced by the media. You can't open a magazine without encountering at least one article about diets, weight loss or dietary supplements intended to make us thinner – with minimum effort, of course. We're bombarded with success stories, and professional models gaze out alluringly from the page. On television, too, the message is clear: we all have to get leaner and fitter and say goodbye to our body fat.

And so it's not surprising that companies put so much effort into selling remedies, dietary supplements, and gadgets like cooling vests (which circulate cold water) to shed our overweight and make our body fat, and thus ourselves, healthier in the process. Although these are eagerly snapped up by consumers anxious to slim down, the results are often disappointing.

But, like everything else in our body, fat is there for a reason. Right? So what actually is fat, and what does it do? And is it really as bad as everyone claims?

Until recently, we knew less about our fat than about any other organ in the body. Yes, fat is an organ, just like our heart and lungs. For many years – for centuries, even – it was thought that fat was little more than a layer of insulation, a blanket of blubber that protects our internal organs from the cold and from bumps and jolts. But nothing could be further from the truth. Research on fat has increased exponentially in recent years, and has provided us with many new insights. If we medical doctors and obesity researchers have learned anything over the past years, it's that our fat is not only one of the largest organs in the body, and an important one, it's an essential one. It provides our other organs with a continuous supply of fuel if we haven't eaten in a while. Our ancestors needed this to survive. But fat does so much more! It also produces countless hormones, substances it releases into the blood to allow it to communicate remotely with other organs, such as the brain. For example, it produces hormones that curb your appetite if you've just eaten a big plate of chips, so that you don't keep eating endlessly. Handy, huh?

There are two sides to our body fat. As long as it remains within reason in terms of size, it will be your friend and keep you healthy. But if you have too little or too much fat, it can turn into a wicked stepsister. When you're underweight, your fat doesn't produce enough of certain important hormones, which leads to a hormonal imbalance that can even result in infertility. And when you're overweight, your fat releases excessive amounts of unhealthy hormones and other substances that interfere with all kinds of bodily processes and can make you sick. Conditions linked to overweight include diabetes (type 2 diabetes, which we will refer to simply as 'diabetes' in the rest of the book), infertility, depression, and also some kinds of cancer.

We feel it's important that you gain knowledge for yourself so you can put this knowledge to good use in your own life to reduce (or increase) your fat mass, make it healthier, or prevent it from inadvertently becoming larger.

In this book, we introduce you to various patients we've encountered in daily practice. Some of them have common body-fat problems (such as overweight or conditions caused by overweight) that will be familiar to many people. Even so, you will see that everyone deals with these problems in their own unique way, and we find this inspiring. We have also decided to tell the stories of a few patients with extremely rare disorders involving their body fat, who have made a lasting impression on us. Over the past few years these special patients have provided science with numerous insights into the amazing way our body fat works, and studying them has led to the deciphering of many of its secrets.

Using both ordinary and extraordinary stories, we take you on a journey through this fascinating organ. How does our body fat actually work? And why does one person have lots of it and another only a little? Do hormonal disruptions have something to do with this? Is too much fat harmful for everyone to the same degree? Why do diets often fail or have only temporary results, and how do you ensure that you maintain a healthy weight over time? Should you avoid stress, stand out in the cold to stimulate fat burning, and make use of other smart strategies? And how do your sleep/wake cycle, your appetite-regulating hormones, and medication affect your fat mass? We all know that unhealthy diet and lack of exercise play a major role in the obesity epidemic. But, in recent years, many other things have also come to light, and it turns out there is an entire world of hidden contributors to overweight! The good news is that you can exert a positive influence on a number of these things yourself, and in this way you can often regain control over your own weight. We'll discuss all of these topics in this book. And we'll give you plenty of practical tips you can use right away. Welcome to the wonderful world of our body fat!

Mariëtte Boon & Liesbeth van Rossum

1

A look at the history of fat

WHY FAT WAS ESSENTIAL TO EVOLUTION

In our modern world there is plenty of food, and you don't have to put much effort into coming up with a week's worth of provisions. You can go to the supermarket on Saturday morning and load up your shopping trolley. Or you can make it even easier on yourself and order your groceries online. But it was very different for our prehistoric ancestors, who had to hunt for their food, and roamed from place to place. This meant that they had to cover many kilometres every day and work very hard, and in spite of this they sometimes still came home empty-handed. Luckily, they had reserves they could always fall back on: their body fat. On days when there was no food available, their body fat would release energy so that important organs like the brain and heart could continue to function. Fat was crucial to survival.

Some of our ancestors were lucky enough to be equipped with an exceptionally efficient energy system. They were able to extract a great deal of energy from only small amounts of food and store this as fat, and they also burned this fat very efficiently. This favourable combination led to more fat reserves, and in times of hunger they could live off these reserves for longer.

So, in prehistoric times, only those people with a substantial amount of body fat could survive harsh conditions such as a prolonged famine. In other words, they had an evolutionary advantage, and this was vital to the continued survival of our species. In prehistory, then, a large amount of body fat was exceptionally prized, perhaps even worshipped – as

1

Figure 1. The Venus of Willendorf.

evidenced by the discovery of a number of mysterious Stone Age statu-ettes. The most well known of these is the Venus of Willendorf (see **Figure 1**), which dates from around 25,000 BC. The statue depicts a woman with a rounded belly, large breasts and broad hips, and is thought to be a fertility symbol. This is ironic, considering severe overweight (obesity) actually leads to reduced fertility.

About 10,000 years ago, following the hunter-gatherer period, a sig-nificant change took place: people settled in one place, the first step in the creation of villages and towns. They kept livestock and cultivated agricultural crops on the land, and could therefore stockpile food. From that moment on, periods of severe famine were less common, although people were still subject to the whims of nature, because crops could fail. For this reason, body fat continued to be an important friend to humans, and this remained the case up until the eighteenth century.

What came next was a period that Robert Fogel, winner of the 1993 Nobel Prize in Economics, called the 'second agricultural revolution'. In his book *The Escape from Hunger and Premature Death, 1700–2100*, he describes how during this period everything changed. In short, it came down to this: improvements in (agricultural) technology led to greater availability of food. This meant that people – who until then had always been small and lean – were able to grow in both height and girth. They became stronger, so had more strength and energy to work even harder, which led to economic growth, new technological developments (e.g. machines), and . . . even greater availability of food. So the population of the Western world ended up in a kind of positive spiral.

But there was a downside to this. At a certain point, people reached the maximum height determined by our genes, while the abundance of food continued. What's more, machines had started doing part of the work, which meant that people didn't have to work as hard. From that moment on, evolution slowly began to turn against us. While using and storing our energy efficiently had once held a great advantage, the wide availability of food and less intensive physical labour meant that people consumed more fuel than they could burn, which resulted in surplus fat reserves. While people had once been small and thin, the streets were now increasingly filled with people who were struggling with overweight and obesity. It took a long time before overweight was seen as a medical problem, which had everything to do with the good reputation fat had long enjoyed.

HOW FAT TRANSFORMED FROM BEING A GREAT FRIEND TO A GREAT ENEMY

The way we view body fat has changed dramatically over the course of history. Just as with hairstyles, what is seen as the ideal amount of body fat is also subject to the whims of fashion. We're all familiar with the voluptuous women with their broad hips and small breasts who graced the paintings of Peter Paul Rubens in the early seventeenth century. This image is now so well recognized that voluptuous women are sometimes said to be Rubenesque.

In ancient Egypt, the streetscape was very different, and the women who strolled along the streets back then were toned and slender. They would make up their eyes with thick lines of kohl, and created complex hair styles. In ancient Greece as well, people (and men in particular) had to be slender and toned. The Greek philosopher Socrates was said to jump up and down every morning to stay slim. According to what we know of the ancient Spartans, fat people were even banned from the city. From the late Renaissance, people wanted to be plump. Just like Peter Paul Rubens, Michelangelo portrayed curvaceous women in his frescos in the Sistine Chapel. Curvaceousness remained popular in the nineteenth century, and was associated with wealth, success and power. This was also not surprising in an age when food was still relatively scarce for many sectors of the population. And when something is in short supply, everything associated with it becomes desirable.

Let's have a look at the early twentieth century. In the tiny town of Wells River, Vermont, in the United States, every year, hordes of men – all of them with large bellies and double chins – would gather together in the local tavern for a weekend of fun. This tavern was the headquarters of the New England Fat Men's Club. Yes, you read that right – this club was created especially for fat men! The club was intended mainly for networking among wealthy businessmen. To qualify for membership, you had to weigh at least 100 kilos and have plenty of money. Influential politicians were also among the members. And the New England Fat Men's Club was by no means the only club of its kind. At the start of the nineteenth century, these kinds of fat men's clubs were springing up everywhere, particularly in the United States but also in France. Fat was in its heyday. Its good reputation was also clearly reflected in the literature of the day. In books by authors such as Charles Dickens, an overweight child was a 'wonderfully fat boy'. Other writers also ascribed positive characteristics such as 'cheerful', 'amiable' and 'good-humoured' to fat people. But that would soon change . . .

Initially, fat's reputation began to decline and become less popular because people simply no longer considered it to be attractive. Around the start of the twentieth century, the ideal was to be slender. From the 1920s, companies hoping to earn piles of money capitalized on this trend. In 1925, cigarette manufacturer Lucky Strike launched a new advertising campaign with the slogan 'Reach for a Lucky instead of a sweet'. Strictly speaking, this works, since the nicotine in cigarettes is an appetite suppressant. Although a cigarette is anything but a good alternative to sweets, of course, it certainly was a clever advertising slogan. In the 1930s, a very successful but also dangerous diet pill came onto the market: dinitrophenol, or DNP. This pill caused body cells to go into combustion mode en masse. Although people shed many kilos, they burned fat at such a high rate that they literally overheated. Some women even died as a result, which led to the pill being taken off the market in 1938. However, eighty years on, it is shocking to note that the pill can still be ordered illegally on the internet. In the 1950s, a new 'miracle drug' came onto the market that was tested by celebrated opera singer Maria Callas, with successful results. She lost more than 30 kilos by taking a pill that contained tapeworm eggs. These grew into long, hungry tapeworms that caused her to lose weight. Although this was certainly effective, it was also both nasty and dangerous. In the 1960s it was fashionable to be thin, and this was further reinforced when the svelte Lesley Hornby ('Twiggy') became a hugely popular model in the UK. Young women in particular wanted to look like her, and not be merely thin, but ultra-thin. The drive to lose weight persisted, and, in 1963, Jean Nidetch – a housewife who said she was obsessed with biscuits – founded the slimming club Weight Watchers. Since then, Weight Watchers has grown into a major slimming empire. Over the past few decades various popular diets have come onto the scene (including Atkins, South Beach and others), and around the turn of the century there were even TV shows where people tried to lose as much weight as possible (to name just two: *The Biggest Loser* and *Obese*). From the beginning of the twentieth century, the desire to be slim was increasingly accompanied by a negative view of people who were overweight or living with obesity. Literature no longer spoke of overweight people in terms of being 'cheerfully chubby' but as 'unsightly fatties'. Also, it was generally accepted that, if you were overweight, it must be your own fault. And that you were 'weak' if you couldn't eat in moderation. Although overindulgence is by no means always the cause, this stigma has huge psychological consequences for many people who struggle with overweight, as we will see in Chapter 11.

Body fat had clearly fallen into disrepute. At the start of the twentieth century this was reinforced when scientific studies conclusively demonstrated that there was an association between obesity and higher mortality rates. It is interesting to note that the first of these studies were conducted by insurance companies. From that moment on, the good reputation fat had enjoyed was lost forever, and from the 1930s it was generally accepted that having too much body fat was a health problem. But how fat impacted health would remain a mystery for a very long time.

THE DISCOVERY OF THE FAT CELL

Let's return to the ancient past. Around the fourth century BC, the Greek physician Hippocrates – who is considered to be the founder of modern medicine – noticed that sudden death was more common among overweight people than among those who were slender. He also noted that obesity was a cause of infertility in women. And he was right, even though he had a different explanation for this. According to him, being overweight made it more difficult to have sexual inter-course, and this made women less fertile. Of course, at that time people were not yet aware of the existence of hormones, let alone the way in which excessive body fat can severely disrupt our hormonal system.

For a long time after, the literature contained little about overweight. It's hard to determine at what point people learned that overweight was the result of excess body fat and not due to an accumulation of another substance in the body, such as blood. There had to have been a moment in history when someone conducting an autopsy (cutting open a body) noticed that a person with obesity had a thicker layer of subcutaneous fat (fat beneath the skin) – which is yellow and spongy – than a person who was slender. It is important to note that, for centuries, until the Renaissance, for ethical and religious reasons it was taboo to cut open human bodies, at least in the Western world. Bodies of the deceased were supposed to remain intact, which is probably why so little was written about this. This changed in the eighteenth century, when count-less books and articles appeared about the causes and consequences of overweight. These included some very interesting theories, and, in retro-spect, some of them were particularly imaginative.

In 1727, Thomas Short wrote that he believed the fat organ to be made up of 'vesicles' (sacs) of fat that were separate from the blood. This was an extremely advanced theory at a time when the entire con-cept of organs being made up of things called 'cells' was still in its infancy. He also believed that overweight was caused by an accumula-tion of both blood and 'oily parts'. According to him, this accumulation resulted from not sweating enough, and so he proposed sweating more as a treatment for overweight. If what he meant by this is that over-weight people should exercise more, he had unwittingly given his patients good advice.

Scottish physiologist Malcolm Flemyng was a student of Dutch sci-entist and researcher Herman Boerhaave in the city of Leiden. Around 1760, he was considering various potential causes of overweight. The first one – that it was caused by consuming too much food, especially fatty food – was right on target. However, he noted that not all over-weight people were by definition big eaters, nor were all slender people necessarily light eaters. Another possible cause was related to the 'fat vesicles' Thomas Short had written of. Flemyng also believed that fat was stored in vesicles enclosed by a membrane. In his view, when these membranes were too limp, it was easier for the vesicles to stretch, and so it was also easier to become overweight. He also wrote that the presence or absence of such 'lax membranes' could run in families. He was thus one of the first to suggest a genetic cause for overweight. Another possible cause of overweight postulated by Flemyng was also in line with the views of Thomas Short: namely, a disruption in the excretion of fluids. He believed that part of the fat in our diet had to be eliminated through sweat, urine and faeces. If this didn't happen to a sufficient degree, your body would store this in fat vesicles and you would become overweight. To remedy this last 'problem', he had also come up with a few good solutions, all of which were aimed at increasing elimination. One of these was rather unpleasant, and involved eating a piece of soap on a daily basis. He described a patient who was said to have lost 14 kilos over the course of two years by eating 2 to 4 grams of soap every day. The theories of Short and Flemyng – that our fat tissue was made up of 'vesicles of fat' – were certainly not ill-conceived. The invention of the microscope by Dutch scientist Antonie van Leeuwenhoek in the seventeenth century made it possible to examine tiny fragments of tissue from organisms, including plants and people, at microscopic level. This eventually resulted in the 'cell theory' (see **Box 1**). After cells were discovered, at the end of the nineteenth century came the discovery of the fat cell as the building block of our fat tissue. And so for a very long time it was generally believed that the fat cell was used for storing fat only, and that together these fat cells formed our fat organ, which enclosed our body in a snug layer and also protected our organs from bumps and jolts.

In the 1990s, this view changed dramatically when fat cells themselves were found to produce hormones, substances that are released into the blood and that can have various effects on other organs. Remotely! And it emerged that fat sends signals to our brain. What's more, fat can have an effect on our behaviour – not just our eating behaviours, but also our mood. Fat suddenly went from being a passive organ to an active one. A new and fascinating field of research was born, research into the many mysteries of fat! Exciting discoveries were made in rapid succession. And every year, hundreds of scientific articles continue to be published, revealing even more of these secrets. But let's start with the basics: how does fat actually work?

Box 1. The cell as building block

All organisms – such as people and plants – are made up of cells. A human being consists of 100,000 billion cells. The cell is the smallest unit within an organism. It consists of a nucleus, in which genetic material (DNA) is stored, and also large numbers of organelles, tiny 'machines' inside the cell that keep it running. An example of an organelle is the mitochondrion, which regulates cellular metabolism. Although all cells have the same 'blueprint', the cells of the various organs can look very different and have completely different characteristics. For example, a muscle cell is very different from a fat cell. Together, cells are the building blocks for tissues, which make up the various organs, such as the heart, the lungs and . . . fat.

2

Fat as an essential organ for storage

Just as a car needs petrol to run, we also need fuel to – quite literally –
move forward. We use a considerable amount of energy every day. Our
heart is constantly pumping blood through our body, we breathe an
average of twelve times a minute, and our liver and kidneys purify our
blood by removing waste products. And that's when we're at rest. When
we exercise, we expend a lot more energy and therefore we need more
fuel. Our body uses basically two kinds of fuel: sugars and fats. In contrast to what many people think, fats are the most important fuel for
most organs. This is because, when they're burned, fats provide the
most energy, much more than sugars. Our body is smart when it comes
to fat, which is why it makes sure we generally have enough of it. It
uses our fat very efficiently. You see, fat is not just our body's most valuable fuel, it also has other essential functions. For example, our body
cells are enclosed in a layer of fat, and fat forms a sheath around our
nerve fibres to enable the nerves to transmit their signals quickly so
that we can think and move quickly. Are you already starting to love
your fat a little?

But where is our fuel actually stored? In a car, this is clear: there's
only one place it can be, and that's in the fuel tank. In our body, though,
the fuel is found in a number of places. A limited amount of fuel – both
fats and sugars – flows freely in the bloodstream, ready to be absorbed
by the cells when needed. This fuel supply in the blood is constantly
being absorbed, and is replenished when you eat. A problem immediately presents itself here. What happens if you don't eat for a while – for
example, when you sleep through the night? Or because there is simply

no food available at the time, as was often the case for our ancestors? Or what if we have eaten but then work out, thus burning more fuel? In all of these cases, we put our stored fuel supplies to good use. They make sure we don't collapse if we skip a meal, and that we can jog or play tennis for an hour with no ill effects – as long as you're not too out of shape, of course. Since our body runs on two kinds of fuel, we also have two kinds of fuel supplies – one for sugar and one for fat – that we can tap into when the fuel in our blood is running low.

GLYCOGEN: STORED SUGAR

Our smallest fuel supply is our stored sugar. This is not sugar in the form of granulated sugar, cane sugar, or beet sugar, but the form of sugar we

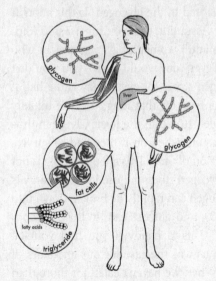 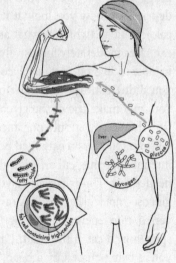

Figure 2. How glycogen and fat is stored in the body.

Figure 3. Using fatty acids and glucose as fuel.

call 'glucose'. To be able to store glucose in the most efficient way possible, it's clumped together into a large ball. These clumps of glucose molecules are called 'glycogen', and are found in two places in the body: in the liver and in the muscles (see **Figure 2**). If the amount of glucose in the blood gets too low (a 'blood sugar dip') – for example, because you haven't eaten anything for some hours – some glucose will be 'clipped' off the glycogen in the liver and released into the blood. Your blood sugar level will then rise again, and you'll be able to keep going. The muscles have their own supply of glycogen, and the glucose released from this supply is used only for the muscles themselves – for example, during vigorous exercise. This is smart, because glucose can be broken down faster than fat, and is thus able to provide energy more quickly.

So exactly how much energy does this glucose supply provide? A total of around 500 grams of glycogen is stored in the liver and muscles. Burning 1 gram of glucose yields 4.1 kilocalories, so a total of between 2,000 and 3,000 kilocalories is stored in the glycogen. Is this a lot? It depends on how you look at it. Assuming that you burn 2,000 kilocalories a day (which is what an adult woman of average weight who exercises moderately intensively every day would burn), you wouldn't be able to live off of your glycogen stores for even a day and a half if you didn't eat. And actually not even for that long, because initially you only make use of the glycogen stored in the liver. Glycogen thus isn't the energy supply our ancestors relied on if they had to survive without food for an extended period of time. This energy supply is not intended for this, either. Your glycogen stores are used when energy is needed fast. This is because glycogen can rapidly be broken down into glucose molecules (see **Figure 3**), and glucose can, in turn, be converted into energy of any sort. This is handy if you find yourself sprinting to catch the train. Or, as was the case for our ancestors, if you're being chased by a tiger. When we haven't eaten for more than three hours or if we have to do something that requires sustained effort, we tap into our body fat.

FAT: A HANDY STOREHOUSE

Our body fat provides us with much-needed energy when no food is available for an extended period of time. And so, over the centuries, our body fat has evolved into a veritable storehouse for energy. It forms early in development, when the foetus is only the size of a walnut, not only under the skin but also in the belly, around the organs. At first, the foetus has only a very modest amount of body fat, and it contains little actual fat. There is no urgent need for it yet, because as long as the foetus is safely inside the womb, the placenta provides fuel through the umbilical cord. However, as the pregnancy reaches its conclusion, the foetus has to be prepared for life outside the womb – it can be cold out there, and it still remains to be seen whether enough nourishment (breast milk) will be available right away. So, to bridge this initial period, at the end of the pregnancy the foetus develops a nice supply of fat. Babies who are born prematurely don't have this fat supply yet, and so they can't keep themselves warm. This means that they have to first spend time in an incubator, a warm and cosy nest in which the newborn needs very little energy to stay at the right temperature. And, over time, the fat store develops on its own.

Our 'body fat organ' is spread out over our entire body, as you may have noticed (or would rather not notice) about yourself. The two largest fat masses are found in the belly around the organs (we also refer to this as 'belly fat') and fat just beneath the skin ('subcutaneous fat'). This subcutaneous fat can be found virtually anywhere, such as in your face (including the familiar 'double chin'), in your feet and in your upper arms. However, to many people's dismay, most subcutaneous fat is found in the abdominal region and in the buttocks and thighs. Why is our body fat so good at storing fat? This is because our body fat is made up of no fewer than 50 billion extremely elastic little balloons known as 'fat cells'. Every one of these fat cells is able to store fat, and can

13

stretch considerably if necessary. When you look at a piece of fat under the microscope, you see that fat cells do in fact resemble tiny balloons – or indeed 'vesicles filled with fat' (see **Figure 2**), just like Short and Flemyng suspected back in the eighteenth century.

But a fat cell is certainly not just a tiny balloon filled with fat – this would by no means do justice to our fat tissue. A fat cell is and does much more than this. For example, a fat cell also has a nucleus and organelles (tiny 'machines') that allow it to make proteins, which makes the fat cell unique. These proteins include the various chemical messengers (in this case, hormones) that make fat such a special organ. And just like every other body cell, fat cells contain special tiny 'power stations' called mitochondria that are responsible for metabolism. We can see the importance of our body fat from the story of Natalie, who turned out to have a rare body-fat disorder.

WHAT IF YOU DON'T HAVE ANY FAT? NATALIE'S STORY

Natalie was eighteen years old, a slender, enterprising young woman from a warm family. But her carefree life came to an abrupt end when her periods became irregular. And that wasn't all. 'I was exhausted, and every move I made was painful. The doctors thought it was glandular fever. But I continued to feel awful, and developed even more symptoms. I could no longer tolerate fatty foods, and was nauseous and vomited a lot. This couldn't all be due to glandular fever, could it?'

Time passed, but Natalie didn't get better. A blood test showed that her blood sugar level was far too high, and at the age of twenty-one she was diagnosed with diabetes. Had this disease been the cause of all of her problems? 'I had to give myself insulin injections, but no matter how much I injected, my blood sugar barely dropped. I was starting to feel desperate. And I was still so tired – I was just barely able to work. It took everything I had, and after work I didn't have the energy to cycle home. The doctors were baffled, and I was at my wits' end. Everyone started to wonder whether there was something more going on here.'

Natalie was referred to a specialist in internal medicine, who noticed that, while her arms and legs were extremely thin, her belly was noticeably rounded. An MRI (magnetic resonance imaging) scan was done, and the results were unusual. 'It turned out I had almost no subcutaneous fat. Instead, there was lots of fat in places it didn't belong, such as on my heart. My liver was also extremely fatty and had become enlarged. That's why my belly was so big! And that's also why I was always so nauseous and why I couldn't tolerate fatty foods. Even though I didn't have any subcutaneous fat, it turned out that the triglycerides [fats] in my blood were very high.'

Natalie was diagnosed with lipodystrophy. This is a rare body-fat disorder that occurs in only one in ten million people. With this condition, for one reason or another, the subcutaneous fat is unable to store fat. Although it contains fat cells, these are virtually empty, and resemble flat, flabby little sacs. This means that the fats have to go somewhere else. They float along in the blood and are deposited at various places – for example, in the belly fat around the organs, which is still able to store fat. A large amount of fat is also stored in other organs – in Natalie's case, this was on her heart and liver, but this could also be around the kidneys. Over time, when fat is stored in these places it can become very dangerous. For example, this can lead to cardiovascular disease, kidney failure and liver disease. Moreover, when fat accumulates in organs, it interferes with the absorption of glucose by these organs and means that the glucose remains in the blood. This causes the blood sugar level to rise, and the person will develop diabetes. As you can see, if your body fat doesn't function properly, the fats and sugars continue to circulate in the blood and ultimately end up in places they don't belong.

Natalie is now thirty-seven, married, and works nearly full-time for a wholesaler. How are things with her at present? 'Right now I'm taking an experimental drug from the United States. Although this won't bring back my subcutaneous fat, it will remove some of the fat from my organs. In the meantime my belly is less swollen, and my sugar metabolism has improved. I also have more energy. I've got my life back.' She even has enough energy for her hobbies – she loves to cycle and enjoys being outdoors. She continues to take the drug from the United States.

Natalie's story shows just how important it is for our body fat to function properly.

YOU CAN LIVE OFF YOUR FAT FOR A LONG TIME

Now we'll take things a step further and have a better look at our fat 'storehouse' – how does fat from our body fat get sent to other organs to be used as fuel? When we talk about the fuel 'fat', which is suspended in our blood and which can be absorbed and burned by organs, what we're referring to are fatty acids. These are long 'tails', or chains, that are usually made up of sixteen to eighteen carbon atoms. Just like glucose, fatty acids are also cleverly packaged so that large numbers of them can be stored without taking up too much space. You can compare it to a zip file on your computer. This is why fatty acids are stored in bunches of three in the form of what are known as 'triglycerides' (see **Figure 2**). Thousands of tightly packed triglycerides are stored in a single fat cell. This is a huge amount of fuel, a real goldmine. As soon as you haven't eaten for a few hours, or if you've been physically active for an extended period of time (for example, exercising or doing housework), this fuel will be tapped into. At this point, fatty acids will be 'clipped' from the triglycerides, released into the blood and then sent to the organs that need fuel (see **Figure 3**). So, in this way, fuel (in the form of fatty acids) is literally 'sent' from your fat to your other organs.

16

Our fat tides us over during times of hunger. But how long can this last? In other words, how long can we live off our fat? Chemistry research studies have shown that burning 1 gram of fat yields 9.4 kilocalories. If you've been paying attention, you will have noticed that this is more than twice as much as the 4.1 kilocalories you get when you burn 1 gram of glucose. This is why our body stores fuel not only as glycogen but preferably in the form of fat. Wouldn't you rather put petrol in your car that lets you drive 700 kilometres instead of just 200? If we wanted to store the same number of calories as glycogen, we would have to lug around many more kilos. And our fat reserves weigh quite a bit as it is. A healthy adult weighing 70 kilos has around 14 kilos of body fat. Since burning 1 gram of fat yields 9.4 kilocalories, this is equal to a total of 131,600 stored kilocalories – an enormous amount! As mentioned earlier, a woman of average build, who exercises moderately intensively every day, burns about 2,000 kilocalories a day, and a man around 2,500 kilocalories. This means that you can live off your fat reserves for an average of 66 days (for a woman) or 53 days (for a man), as long as you don't become more active.

In practice, you would actually be able to manage for even longer, because we have a third energy supply, one we would rather not tap into unless our life depends on it. This is our protein reserves. A healthy adult who weighs 70 kilos is carrying around nearly 10 kilos of protein, half of which can be broken down into tiny pieces known as amino acids, which are then used as fuel. Your body would rather not touch this energy supply, because the body's own proteins are not intended to be used as a source of energy. These proteins are an important constituent of muscle, including the heart muscle and respiratory muscles, and also play a role in defending the body against pathogens. This becomes clear in patients with the disease anorexia nervosa, who eat very little or nothing at all for an extended period of time. Along with having very little body fat, a person with anorexia also has very little muscle, because the body has broken this down to get enough energy. Moreover, even

17

the most innocuous viruses, bacteria and funguses can make someone with extreme anorexia sick because their immune system can no longer fight them.

FAT LOVES SUGAR – AND SO DO WE

Eating is one of the major ways we replenish our energy supplies. And that works out well, because most people love to eat. So do we, by the way. The food you eat contains three different building blocks: sugars, fats and proteins. Before our food can be absorbed by the gut, it first needs to be broken down into smaller pieces. This is aided by digestive enzymes, which can be seen as tiny 'scissors' that snip large pieces of food into small building blocks.

To start with, let's have a better look at the building block 'sugar'. 'Sugar' often refers to carbohydrates that consist of just one or two building blocks, and are generally known as 'simple carbohydrates'. An example of this is granulated sugar. Most sugars are made up of glucose, fructose, or a combination of the two. Fructose occurs naturally in foods like fruit and honey, and because of this is also called 'fruit sugar'. Because fructose is so sweet, it's also used as a sweetener in a wide range of products, such as biscuits, chocolate and sweets. There are also carbohydrates, for instance starch, that consist of long chains of glucose. This type of carbohydrate is found in foods like bread, potatoes and pasta. Furthermore, there are carbohydrates we call 'complex carbohydrates'. These complex carbohydrates form fibres, and occur along with fibre in such things as vegetables, wholewheat pasta and bread, brown rice, lentils, fruit, nuts, legumes and seeds. Complex carbohydrates are generally the healthier form of carbohydrate.

Glucose and fructose (the simple carbohydrates) can be absorbed directly by the gut. The complex carbohydrates first have to be 'snipped' into individual sugars by digestive enzymes before they can be absorbed by the gut. The more complex the structure of the complex carbohydrates (for example, when there are many side branches), the more time and effort it takes for the enzymes in your gut to snip them into pieces. After you eat a food that contains many simple carbohydrates (so, one that contains individual sugars), your blood sugar level will rise much faster than it would after eating a meal containing complex carbohydrates

---

Box 2. The difference between simple and complex carbohydrates

Simple and complex carbohydrates differ not only in their make-up, but also in the effect they have on our body. The table below contains an overview:

	Simple carbohydrates	Complex carbohydrates
Types of foods	White bread, white rice, crisps, sweets	Wholegrain bread, brown rice, wholewheat pasta, vegetables, legumes
Effects on the body	Rapid rise in blood sugar level, which will boost your energy quickly	Slower and less pronounced rise in blood sugar level, which will boost energy later on
Effects on insulin	Rapid rise and high peak insulin level. This also results in a rapid drop in blood sugar level and a 'tired' feeling	Slower rise and lower peak insulin level. This means that the blood sugar level stays higher longer, and the boost in energy will last longer

---

(see **Box 2**). When you get hungry, there's a reason you might quickly feel the urge to grab something that will make your blood sugar level rise rapidly, like a piece of chocolate. This will boost your energy fast.

When your blood sugar level rises, this signals the pancreas to release an important hormone: insulin. Insulin is a kind of guide that shows all of the glucose that enters the blood where it needs to go. Insulin shows the way by opening doors in the various body cells so that the glucose can enter and the cells can then use this as fuel. And both muscle and the liver can convert the glucose to glycogen to replenish the sugar supply. But insulin does much more! If more glucose enters the blood than can be used immediately or converted into glycogen, there is a surplus. This surplus is absorbed by our body fat and converted into triglycerides – in other words, into fat! So a prolonged surplus of sugary or starchy foods – whether soft drinks, biscuits, white rice or potatoes – can ultimately lead to overweight in the same way as an excess of foods high in fat. This doesn't mean you have to eliminate all carbohydrates from your diet. We need carbohydrates as fuel for the brain and to build up glycogen reserves. Furthermore, the fibres are important to keep our gut healthy, and people who eat lots of fibre have a lower risk of developing diabetes and cardio-vascular disease. But moderation is key!

Nowadays, though, it has become very tricky to do things in moderation. We all know there is sugar in biscuits, cakes and other sweet things. But did you know there is sugar in many processed foods? You'll find it in packaged soup, bread, tomato sauce, and even in sliced meats. We call these 'hidden sugars'. It turns out that manufacturers hide sugars by using other names on food labels. When we see 'vanilla sugar', we still think of sugar. When we see honey or syrup on a label, sugar may not necessarily be our first thought. But what about 'cassonade' and 'rapadura'? They sound more like exotic dances. But no – these are both different words for sugar. That's what you might call a 'sweet surprise'. Why do manufacturers put sugar into everything? This is because we have an inborn love of sweetness. For example, mother's milk is wonderfully sweet. When extra sugar is added to products, we find them even more delicious, and so we'll keep on buying – and eating – them. Manufacturers are well aware that we find products without sugar less tasty.

WHY DO WE GET GRUMPY
WHEN WE'RE HUNGRY?

Does the following situation sound familiar? Miranda came home from work hungry one day. She'd had lunch already at half past eleven because she had back-to-back meetings all afternoon, and hadn't got around to eating her usual afternoon apple. When Miranda got home at half past six, her husband Patrick noticed she was excessively annoyed that he'd left his shoes lying around, and she muttered angrily about them under her breath. Her hands were trembling. Patrick offered to start making dinner right away. She'd planned a meal of chicken breast, vegetables and basmati rice. But when Miranda went into the kitchen, she saw he was making risotto with large chicken legs instead – and that it was nowhere near being done. She exploded, shouting at him that he should have talked with her first about what he was going to cook. And why on earth did he decide to make something that was going to take 45 minutes to cook instead of the quick meal she longed for? What was he thinking! She grabbed a bag of prawn crackers and a banana and stomped upstairs. She flung herself onto the bed and wolfed down her snacks. Before long she felt better. Her hands were no longer shaking, and she felt calmer. She gradually realized she'd overreacted, and, after half an hour, went downstairs and apologized to Patrick. Patrick smiled. He knew her well, and had seen it coming. Calmly, he handed her a now lukewarm chicken leg.

There's a wonderful word for this: 'hanger', a combination of the words 'hunger' and 'anger'. Is hanger real? Or is this nothing more than an invented excuse for bad behaviour and emotional outbursts that, uh, women in particular are guilty of? No. This really is based on science. If you haven't eaten for some time, your blood sugar level drops. In response, the body releases various hormones that enable the blood sugar level to rise again. This is very important for the brain, because glucose is the brain's main fuel. If the blood sugar level gets too low, this could even cause damage to the brain. Two of the hormones that cause the blood sugar level to rise are adrenaline and cortisol. These in turn also have their own effect on the brain. These hormones enable the brain to produce neuropeptides, tiny 'messenger proteins' that transmit signals. And it turns out that these neuropeptides (such as neuropeptide Y) also activate feelings of anger, irritation and impulsive behaviour in the brain. So it's not surprising, then, that when someone is hungry they're also much more likely to react with irritation – and perhaps also take decisions more impulsively. So maybe from now on you shouldn't do any online shopping or ask your boss for a raise if you're hungry . . .

The best way to prevent the kind of low blood sugar that can make you grumpy is to eat healthy foods that are also filling: wholegrain products, fresh fruits and vegetables, a handful of unsalted nuts, and yoghurt, which contain complex carbohydrates, lots of protein, and unsaturated fats. It's best to avoid beverages high in sugar (see **Box 3**). Another way to keep from having low blood sugar was discovered by the American medical doctor and researcher Alpana Shukla at the Comprehensive Weight Control Center in New York. For many years, she has been conducting intensive research into obesity and weight loss, and recently studied whether the order in which you eat the various parts of your meal makes a difference. And what emerged? If you eat the protein-rich foods first (such as your egg, yoghurt or milk) and only then the carbohydrates (such as your rice or bread), after the meal your blood sugar level will be lower than if you had eaten the carbohydrates first. You can use this clever trick if you have diabetes and need to keep your blood sugars under control. Even if you don't have diabetes, it may be helpful to keep you from getting hangry later in the day.

Box 3. It's better to eat your fruit than to drink it!

Is your blood sugar low and are you about to chug down some freshly squeezed fruit juice? It's better not to make a daily habit of this, because it's not as healthy as you think. Food in liquid form (such as orange juice) often goes from the stomach to the first part of the small intestine in ten minutes flat, while this takes more than an hour for solid food (such as an orange). When we drink fruit juice, we can gulp down a large quantity of sugars in no time. One glass of fresh orange juice contains the same amount of fruit sugars as three or four oranges. If you were to eat these instead (and usually you would eat only one or two, not four!), you would also take in all kinds of other nutrients at the same time – especially fibre, which your body uses energy to digest. The fibre also causes the fruit sugars to be released much more gradually, thus avoiding an undesirable blood sugar spike. Solid food is more filling and satiating.

And if you can't resist smoothies, then add some banana or Greek yoghurt. These will thicken the structure of your smoothie so it will remain in your gastrointestinal tract longer, which gives your body more time to produce satiety hormones. Yoghurt is particularly filling because of the protein it contains, and has relatively few calories.

When it comes to products that provide the greatest number of calories in the shortest amount of time, the absolute front runner is chocolate milk. This is due to the combination of high-calorie fats and sugars along with the fact that it's a liquid.

DIGESTING A HIGH-FAT MEAL

Now we'll look at how our body digests the fats contained in a meal. As previously mentioned, fats are made up of fatty acids stored in bunches of three, which we call triglycerides. A high-fat meal can be made up of many different types of fat, and these differences are found in the kinds of fatty acids in the triglycerides. To start with, fatty acids can have different lengths. And you've probably heard of 'saturated' and 'unsaturated' fatty acids. The chemical structure of these fatty acids is different. Unsaturated fatty acids are found in things like fruits and seeds, such as olives and flaxseed, and also in certain kinds of fish (such as mackerel and salmon) and nuts. High-fat food products that are soft or liquid at room temperature (such as oils) also contain lots of unsaturated fat. Saturated fat is found in many animal products, like meat and high-fat cheese. Unsaturated fat is generally considered to be 'healthy fat' because if your diet contains plenty of this kind of fat, you have a lower risk of cardiovascular disease.

Say that one night you indulge in a meal that's high in fat. After the fat ends up in your stomach, it still has a long way to go. First, it is 'snipped' into individual (saturated/unsaturated) fatty acids in the gut. Then, before entering the bloodstream, these fatty acids undergo a major change. This is necessary because fats aren't water-soluble. Just think of what happens when you add a little olive oil to the pan when you're cooking pasta – the water and oil don't mix. This is exactly what would happen in the blood if fats weren't packed into special water-soluble containers. These are large fat-rich globules that transport huge quantities of fats such as triglycerides in the blood, where they can be delivered to the organs in need of extra energy, particularly to muscle and fat tissue.

The hormone insulin – which the pancreas releases into the blood when the blood sugar level rises after a meal – is involved not only with distributing glucose throughout the body, but also fats. Insulin activates a certain protein in various organs, and this protein can 'snip' the fatty acids from the fat-rich containers in the blood. The relevant organ then absorbs these fatty acids from the blood. Our fat tissue also absorbs large quantities of fatty acids in this way. And on top of this, insulin also causes fewer fats to be broken down in the fat tissue. Which results in more fat being stored!

So the hormone insulin can be seen as a real helper to our fat tissue, and from an evolutionary point of view it's a true survival hormone. It uses many strategies to promote the storage of fat. This is one of the reasons why people who need to inject a large amount of insulin to treat their diabetes find it even harder to lose weight. Insulin holds tightly onto fat! For patients who inject insulin and want to lose weight, the amount of insulin they inject is lowered just before making a life-style change (such as eating less and exercising more) to prevent their blood sugar from dropping too low. Lowering the amount of insulin also helps them to lose fat! Of course, this must always be done in consultation with a doctor.

WHEN DO YOU HAVE TOO MUCH FAT MASS?

In today's world – a world in which food, and particularly unhealthy food, is literally at our fingertips – there is an ever-present risk of developing too much body fat. But how much is too much body fat? A quick and easy way to determine this is by calculating your body mass index (BMI) (see **Box 4**). BMI is synonymous with the Quetelet index, which was first described by the Belgian Adolphe Quetelet. He was an impassioned mathematician and statistician who wanted to capture human beings, to the greatest possible extent, in statistical models. In 1832, he conducted research into the average ratios between height and weight in humans, and showed that weight increases as a square of the height (BMI = weight/height^2, whereby weight is in kilos and height is in metres). In the twentieth century, overweight was increasingly seen as a medical problem, and reference ranges for BMI were established based on major studies.

A BMI of between 18.5 and 25 is considered to be a healthy weight. A BMI of between 25 and 30 indicates overweight, a BMI over 30 indicates obesity or severe overweight, and a BMI over 40 indicates morbid obesity.

Box 4. What is overweight or obesity?

The word 'obesity' comes from the Latin *obesus*, and literally means 'overeating' (ob = over, esus = eating). This term was first described in the literature in 1611. Overweight and obesity are determined using the ratio between your height and weight, which is known as the body mass index (BMI). Your BMI is calculated as weight (in kilos): height^2 (in metres) and expressed in kg/m^2. BMI is divided into the following categories (for both men and women):

BMI < 18.5 = underweight
BMI 18.5–24.9 = healthy weight
BMI 25–29.9 = overweight
BMI 30–34.9 = obesity (class I)
BMI 35–39.9 = obesity (class II)
BMI ≥ 40 = obesity (class III or morbid (severe) obesity)
BMI ≥ 50 = super obesity

A distinction can also be made based on waist circumference. In men, a waist circumference greater than 94 cm indicates overweight, and a circumference greater than 102 cm obesity. In women, a waist circumference greater than 80 cm indicates overweight, and a circumference greater than 88 cm obesity.

For people of South Asian descent, lower thresholds apply to both BMI and waist circumference.

Slightly lower thresholds apply to people of South Asian descent because they are generally more likely to develop complications such as diabetes at a lower BMI. Although BMI is a handy way to quickly calculate whether a person is in the 'safe zone', it can sometimes be misleading – for example, if a person does bodybuilding and has lots of muscle mass. Since muscle is heavier than fat, such a person can have a high BMI even though there is a healthy amount of body fat. This is why the body fat percentage (which you can measure with special scans, special scales, or a skinfold measurement) is a better measure of the amount of body fat. In general, the body fat percentage is too high if this is over 20 per cent for men, and over 30 per cent for women. And no, these thresholds weren't thought up by a woman and therefore are less strict for women! Female sex hormones play an important role in this difference.

But body fat percentage doesn't tell the whole story. It tells us nothing about where the excess body fat is located. Body fat located in the belly around the organs is more harmful than subcutaneous body fat (body fat under the skin). Measuring waist circumference will provide an indication of this. A healthy waist circumference is between 74 and 94 centimetres for men, and between 68 and 80 centimetres for women. Here as well, lower guidelines apply to those of South Asian descent. Unfortunately, because waist circumference is not routinely measured in the consultation room, we (doctors, nurses and also the policy makers) still use BMI as an indicator of overweight. If we take BMI as our frame of reference, according to the World Health Organization (WHO), worldwide 39 per cent of adult people have overweight and 13 per cent of people are living with obesity. Among children aged four to nineteen years, 18 per cent are overweight. In the United States these figures are even higher, and, by 2030, 86 per cent of adults are expected to have overweight or obesity and 51 per cent obesity.

OVERWEIGHT: THE RESULT OF A POSITIVE ENERGY BALANCE

Overweight is associated with various health risks. Therefore, a great deal of research is currently being funded to find new ways (such as drugs) to combat overweight. We will go into this in more detail in Chapter 10. But, to be able to tackle overweight effectively, we first need to understand exactly how it develops. The basics seem very simple: you gain weight (and ultimately become overweight) when, over an extended period of time, you take in more fuel than you burn (for instance, if you eat one biscuit too many every day for a number of years, this can ultimately lead to overweight). But it's not that simple. It turns out that there are many factors that play a role in causing us to gain weight or in making it virtually impossible to lose weight. And this is why one person is much more prone to becoming overweight than another. Who doesn't have a friend who always seems to be on a diet but just can't manage to lose weight? And how is it possible that some people seem to be able to eat everything and rarely exercise but still stay as thin as a rail? It should be clear by now that the origins of overweight are very complex. As we shall see in various chapters further along in this book, there are countless factors that have an effect on our fat. In addition to insulin, there are all kinds of other hormones as well, such as the thyroid hormone, the stress hormone cortisol, and sex hormones like oestrogen and testosterone. And this is just a small selection of the hormones that can affect our fat.

The brain also plays a role. Our brain, which makes countless connections with the rest of our body. Which enables us to speak, chew, cycle, laugh and cry. Which determines our mood. This same brain also has close ties with our fat, via nerves that provide a direct line of communication with it. There are special areas of the brain that register whether the rest of the body needs extra fuel. If this is the case, the brain can tap into our fat straight away and get the fat to release fatty acids into the blood. The brain can also make you feel hungrier if your body doesn't have enough fuel. In some people these areas of the brain are damaged. As a result, the fat tissue holds on to the stored triglycerides as best it can and the body fat is hardly used, or people always feel extremely hungry. This kind of damage will always lead to overweight. So, in contrast to what some people believe, overweight is not always simply a matter of eating too much or not exercising enough. But sadly, the stigma that overweight is always a person's own fault, and that this must mean they're weak for letting things go so far, still exists.

WHAT DOES A HEALTHY DIET LOOK LIKE?

Carbohydrates, proteins, fats: in recent years, it seems more books have been written about 'effective weight-loss diets' and 'healthy diets' than about any other subject. So, should we be eating lots of fat and very few carbohydrates? Or lots of carbohydrates and very little fat? We all seem to have lost our way to some extent.

A little bit of history: in the 1970s and '80s, governments strongly encouraged people to avoid fats as much as possible. Excessive fat consumption was thought to be to blame for overweight, diabetes and cancer. Over the past few years, another trend entirely has emerged: carbohydrates, and especially sugar, not fats, are the culprits. So now the focus is on encouraging people to eliminate carbohydrates from their diet. What should you do?

Let's start with this somewhat unsatisfactory conclusion: there is no such thing as a single optimal diet for everyone – because everyone is different. In the future, we will likely move towards personalized diets where it will be possible to tailor the quantities and kinds of fats, carbohydrates and proteins to you as an individual, even taking your genetic make-up into account! But we're not quite there yet. We see that people sometimes sing the praises of certain diets with something approaching religious fervour. If a diet has worked well for them or their neighbour, they're convinced it will work for everyone. In practice, some people will indeed have benefited from it, and some will have benefited from a completely different diet, but for some, a diet won't have worked at all.

Recently, researchers at Stanford University in the United States published a major scientific nutrition study that provides us with more insight into this. They randomly assigned 609 men and women with overweight or obesity without diabetes to go on either a very low-carbohydrate or very low-fat diet. Over the course of twelve months, the participants attended twenty-two group sessions that focused on the diet they had been assigned to, exercise, emotional wellbeing and behaviour change. In the end, it turned out there was no difference in terms of weight loss between the two diet groups: those in both the low-carb and the low-fat groups lost 5 to 6 kilos on average. What was striking were the individual differences. In both diet groups, some people lost as many as 30 kilos, while others gained 10 kilos! These kinds of studies show that the details of the diet don't really matter that much as long as the basic diet is healthy. Both groups followed a diet that consisted of as many vegetables as possible, and as few added sugars, refined flour products, and processed foods as possible.

What we do know is that certain diseases respond much more favourably to specific diets. People with diabetes or prediabetes (less effective sugar metabolism) can benefit from a low-carbohydrate diet with good levels of fat and protein, precisely because these people are less able to utilize carbohydrates efficiently. But, as already mentioned, the kind of fat you eat also matters. Saturated fat can elevate LDL cholesterol, and this in turn can increase the risk of cardiovascular disease. This is why people are advised to replace foods high in saturated fat with foods high in unsaturated fat. This is in line with the general dietary recommendations of the World Health Organization (WHO). In general, eat plenty of vegetables and fruit every day. Also eat fat every day, and this should be unsaturated fat for the most part. Choose white meat (such as poultry) as well as oily fish (like mackerel or salmon) once a week. Eat only limited amounts of processed meat, since these are high in fat and salt. Regularly eat a handful of unsalted nuts and/or some legumes. Eat or drink dairy products every day, because they contain proteins and other beneficial nutrients. Choose wholegrain bread, rice and pasta. In terms of beverages, try to limit the number of glasses of sugary drinks (also 'diet' drinks) as much as possible, and drink lots of water, and also tea and coffee without sugar. And drink alcoholic beverages sparingly. You can find more information at the WHO website '5 keys to a healthy diet' (https://www.who.int/nutrition/topics/5keys_healthydiet/en/). If you don't consume more calories than you burn on a daily basis, there's a good chance that this advice will have a positive effect on your weight and on your metabolism.

3

Fat as a hormone factory

THE GIRL WHO IS ALWAYS HUNGRY: KAREN'S STORY

Karen is a cheerful six-year-old from a happy family that includes her sister and her parents Ilonka and Kevin. She likes games and playing outside. However, Karen is the only one in her family who is overweight. She weighs 40 kilos, while a child of six should only

Karen (on the left), who is four years old in this picture, is always hungry and keeps gaining weight.

weigh half this, and she's always hungry. Soon after Karen was born, her mother Ilonka noticed that something wasn't right with her daughter. 'Karen cried a lot as a baby, without any clear cause. She was only quiet once she'd had something to eat, but then a short time later she'd start again. We often felt desperate. Although we were careful not to give her too much food, Karen gained weight fast, sometimes 4 kilos a month! People would often say she looked "well fed". At a certain point I couldn't stand to hear that expression any more. I felt like there was something wrong with Karen. It just didn't seem possible she could gain weight so fast from the limited amount of food we were giving her. At a certain point I was going to the Well Baby Clinic every week. I was afraid she was going to burst out of her skin! I kept a diary in which I wrote down exactly how much she ate and when she ate it.'

When Karen was six months old, Ilonka showed the diary to the doctors. They were so alarmed by her weight that they were referred to a university teaching hospital. 'I so hoped they would be able to find an explanation for this. And better yet, that there would be a treatment available for Karen! But unfortunately things didn't turn out to be that simple. Karen was first tested for the most common causes of childhood obesity, such as an underactive thyroid. It turned out she had none of these. Then genetic material taken from Karen was tested for rarer causes of obesity, and again, nothing was found. I started to get desperate. By now Karen was one and a half and already weighed 26 kilos, even though she was on a diet drawn up by a dietitian. That diet consisted of a limited number of calories that Karen would get in small portions every two or three hours.'

Although Ilonka followed her daughter's diet to the letter, it wasn't easy, and during this time they were having difficulty dealing with Karen. 'Karen would be content just after she ate, but then she would have tantrums. She wanted more food! I found that terribly difficult – I didn't give it to her, for her own good, but as a mother, it was heart-wrenching. In the meantime, I was having an increasingly hard time. Every day when I went out with her for a walk, I was confronted by the looks we got from the people we passed. I knew what they were thinking! And sometimes they would even say things out loud. There was the time Karen was eating a piece of cucumber, and I heard someone mutter, "Go ahead, just stuff the child full of food." Or people would whisper to each other that I must be a bad mother and that the Child Welfare Office should get involved. It's bizarre, all the things people have the nerve to say. At first I would defend myself, but after a while I just couldn't bring myself to do it any more. As a result, there was a time when I barely left the house. And I started to doubt myself: maybe it really was my fault.'

When Karen was two, she was hospitalized for two weeks to assess whether her obesity had resulted in any apparent consequences thus far. Fortunately, she didn't have diabetes, but she did have some fat deposits on her liver. The cholesterol level in her blood was also too high. 'Then we were even more motivated to get Karen as healthy as we could, but what this meant in practice was mainly depriving her of food. We got more and more creative in how we made it clear to Karen when she could eat and what. For example, we made a clock with pictograms on the various numbers that showed the food she could have then, like fruit or a sandwich. This calmed her down somewhat. In the meantime, the testing for a possible cause continued. Then, when Karen was two and a half, we got the call we'd been hoping for. Clinical geneticist Mieke van Haelst and paediatric endocrinologist Erica van den Akker had found the cause: Karen was missing the receptor, or receiver, for the hormone leptin, which meant that her brain was constantly sending a signal telling her she was hungry!'

THE DISCOVERY OF THE FAT HORMONE LEPTIN

What emerged was that Karen's extreme obesity – which had turned her life and the life of her family upside down – was caused by a single 'mistake', or mutation, in her genetic material: her DNA. This is called 'monogenic obesity'. Our DNA contains the code for all of the proteins manufactured in our body, such as the proteins that together make up our muscles or eyes. Mutations in the DNA can result in deformed proteins and thus lead to diseases, such as certain muscle diseases. Karen's mutation was right at the spot (so, in the gene) that was supposed to produce the protein for the receptor (the receiver) of the hormone leptin. In Karen's case this mutation led not only to an increased appetite, but also to a lower resting metabolism. A very difficult combination, since Karen is extremely hungry, but can eat less calories than average.

Don't get the idea that everyone who has a large appetite has this mutation, however. In fact, this abnormality is so rare that right now there are only six children in the Netherlands who are known to have a mutation in this gene. Worldwide the total number of published patients with a gene mutation in the leptin receptor is eighty-eight, of which twenty-one are European. Together with obesity researchers Lotte Kleinendorst and Ozair Abawi we recently studied over 77,000 European individuals and calculated that at least 998 patients with a defect leptin receptor should be present in Europe. So, probably, the majority of patients with a defect leptin receptor are never diagnosed!

Although it may be rare, this DNA mutation and its consequences have taught us a great deal about how our fat works. It has even led to ground-breaking new discoveries. The long road of research that led to these discoveries is so interesting that we would like to go into this in greater detail here.

To do so, we need to return to the 1940s. At that time, a group of scientists was breeding mice for animal experiments in a large laboratory in the American town of Bar Harbor, Maine. These mice were genetically identical, and thus had the same DNA, just like identical twins. As a result, during medical experiments – for example, when testing a drug – the mice respond in virtually the same way to the drug being tested. So any differences in the test results can easily demonstrate the effects of the drug.

But in the summer of 1949, the laboratory staff noticed something strange. A litter of mice was born that was much heavier than the other mice from a young age. Further observation revealed that these mice were much less active than the leaner mice, and that they ate more – much more. One of the mice was so hungry that he lay sprawled out in his cage with his head in his food bowl so he could go on eating all day long. Something strange was going on here! The laboratory staff reasoned that a mutation must have occurred in the mice's genetic material, and decided to compare the DNA of the fat mice with that of their lean parents. And sure enough, in one particular location, the DNA of the fat mice was different from that of the lean mice. A mutation had occurred here, and was given the name Ob (for 'obesity'). This DNA mutation proved to be incredibly important to obesity research. Until then – and this was now the early 1950s – little was known about how obesity developed. Would this unravel one of its mysteries? The researchers were raring to go. It so happened that other researchers had just discovered that a fascinating area in the brain known as the hypothalamus (which we will discuss in greater detail in Chapter 5) seemed to be responsible for a feeling of fullness, or satiety, following a meal. When this area in the brain was damaged in mice, they became obese because they no longer ever felt full and so were always hungry. During this same period, British researchers Kennedy and Hervey hypothesized that fat tissue might be responsible for producing a hormone (see **Box 5**) that was released into the blood and that, by binding to the satiety centre in the brain, was responsible for satiety. This was a cutting-edge theory, because until then it had always been thought that the only thing our body fat did was store fat. Could the discovery of the obesity gene and this hypothesis be related? It was worth investigating this further, all the more because a short time later – in the same lab in Bar Harbor – yet another litter of mice from a different strain was born that became obese early on and had voracious appetites. The difference with the Ob mouse strain was that these mice developed diabetes from an early age. Therefore, the gene in which this DNA mutation was located was given the name Db (for 'diabetes').

Box 5. What is a hormone?

A hormone is a chemical messenger that is released into the blood by a hormone gland and which produces effects further along in the body. Hormones do this by binding to what are known as hormone receptors on the target organs. This triggers various reactions within the organ, such as increased fat burning. A hormone fits into a receptor the way a key fits into a lock. The thyroid gland, the adrenal glands and the hypophysis or pituitary gland all produce hormones. Thyroid hormone is one example of a hormone. But hormones can also be produced and secreted by other organs, such as the heart and . . . our body fat!

Then, at the end of the 1960s, it was time to thoroughly investigate what the Ob and Db genes did exactly. Since much of the modern technology we now have at our disposal did not exist at that time, an experiment had to be devised that would be able to determine the effects of the mutations in the Ob and Db genes. This led to the following rather gruesome experiments. First, the blood supplies of two mice – a 'normal' mouse and an 'Ob' mouse (from the Ob mouse strain) – were connected. This radical technique that joins two living beings is called parabiosis – both organisms share a common blood supply, just like conjoined twins. What happened next was utterly fascinating. The Ob mouse began to eat less and less and lost weight, until it was just as lean as the normal mouse it was attached to!

A number of conclusions can be drawn from this experiment. Apparently, the Ob mouse's blood did not contain the substance that produces satiety. However, when this mouse was surgically joined to the normal mouse and their bloodstreams were linked, this changed. Evidently the normal mouse was producing this substance, and this was being passed on to the Ob mouse through the blood. As a result, the Ob mouse lost a huge amount of weight from the moment it was attached to the normal mouse.

There was also a second experiment, in which the blood supply of a 'Db' mouse (from the Db mouse strain) was connected to that of a normal mouse. Now something very different happened! The normal mouse rapidly lost weight, and died of starvation within fifty days. We will explain why this happened in more detail shortly. However, the Db mouse was unaffected – it did not lose weight and continued to eat large quantities of food. This suggests that, unlike the Ob mouse, the Db mouse was resistant to the mysterious substance in the normal mouse's blood.

It wasn't until 1994 that it became clear exactly what this substance was and where it was being produced in the normal mouse. It concerned a hormone that was being produced in large quantities by the normal mouse's fat, and which wasn't being made at all by the Ob mouse's fat – which meant that the Ob mouse lacked this hormone. Subsequent experiments showed that when Ob mice were injected with this hormone, they ate less and had less body fat. You could say they had been 'cured' of the condition that had developed as a result of their DNA mutation. This hormone was given the name 'leptin', from the Greek leptos, which means 'thin'. With its discovery, leptin was the first hormone shown to be produced by our body fat.

But what does that piece of DNA that contains the Db gene do? This forms the code for making the leptin receptor. As we explained earlier, a hormone can only do what it's meant to do if it can bind to its receptor (see **Box 5**) – in the same way that the front door to your house will only open if the key fits the lock. The same is true for leptin. These leptin receptors can be found in a number of places in the body, including the satiety centre in the brain. The Db mouse has no problem making leptin – in fact, it actually makes lots of it because it has so much body fat. The problem is that this mouse's leptin receptor is defective, and so the leptin isn't able to produce a sense of satiety. This explains why, in the experiment in which the Db mouse was joined to the 'normal' mouse with functioning leptin receptors, this normal mouse died shortly after being joined to the Db mouse. The Db mouse had extremely high levels of leptin in its blood, which decreased the appetite of the normal mouse to such an extent that it starved. So, as you can see, paying close attention while breeding mice can lead to ground-breaking discoveries: leptin turned out to be a determining factor for the feeling of satiety.

LEPTIN IN HUMANS

Following the mouse experiments, of course the next question that arose was what the outcome of the studies could mean for humans. Shortly after the discovery of leptin, two extremely obese Pakistani children from the same family were studied. Just like Karen and the obese mice, these children had been insatiably hungry and severely overweight from a young age. It turned out that the leptin levels in their blood were nearly undetectable and, just like the American mouse, they had a rare mutation in their Ob gene, so that their fat could not produce leptin. This finally explained all the years of extreme hunger and struggling in vain to lose weight.

For obesity researchers, these were exciting and inspiring times. In 1998, as budding researchers, one of us attended a conference of the American Diabetes Association in Chicago. At the conference, Professor Stephen O'Rahilly of the University of Cambridge in the United Kingdom, one of the lead researchers, talked about the leptin deficiency that had been found in the two Pakistani children with extreme obesity. O'Rahilly concluded by announcing that they were now treating the oldest of the two children, a nine-year-old girl, with synthetic leptin. Although at the time he wasn't free to discuss the results, his mischievous smile gave rise to high expectations. The atmosphere in the hall of hopeful researchers turned electric. Had a treatment for this rare form of obesity finally been found?

A year later, the answer was revealed: yes, leptin therapy worked! In an article they published in the prestigious *New England Journal of Medicine*, O'Rahilly and his research team described how the girl lost more than 16 kilos in a year, whereas before she had only gained weight. This was revolutionary news in the obesity world: not only a mouse but also a human with morbid obesity caused by a faulty gene could be successfully treated with leptin.

Other patients that could not produce their own leptin (known as 'leptin deficiency') were also ecstatic when leptin proved to be effective. Shortly after the start of treatment, they noticed major differences: their constant hunger vanished, they ate less, and the kilos melted away. After a couple of years, they were unrecognizable, and their weight was nearly normal. Around the world, more than thirty people who suffer from leptin deficiency have now been successfully treated with leptin. They have been spared a lifetime of obesity and a premature death from related complications.

In Karen's case, it turned out to be something else in the end. The genetic tests showed that, although she was able to produce normal quantities of leptin, she was missing the leptin receptor – just like the Db mouse. Although outwardly this looks nearly the same as when a person is missing the leptin itself (namely, extreme hunger and obesity), there is one important difference. Her mother was all too aware of this. When the leptin receptor is missing, treatment with synthetic leptin is of no use because there is no receptor. No treatment is yet available for patients like Karen, other than to try to eat as little as possible and fight the hunger.

There are some promising initial results with new drugs that can bypass the leptin receptor and make use of another receptor to produce a feeling of satiety. Time will tell whether these drugs will be able to help Karen.

The past few years have had their ups and downs: 'Karen is a little girl who listens very well. She knows all too well what she can and can't eat. At birthday parties she dutifully eats an apple while the rest eat sacks of crisps. Once in a while she'll still get upset if she doesn't get extra food between meals. We also come up against lots of practical things in daily life. Karen's shoes have to be specially made because her feet are so big. Normal jackets don't fit either – I have to buy them a couple of sizes larger, and then have the sleeves shortened. Karen would also dearly love to wear pretty dresses, just like her sister, but they simply don't fit her because her belly is so large. Recently, she found herself in a very distressing situation. We were at an amusement park and Karen got into one of the little cars, just like the other children. When the ride was over, everybody had to quickly get out of the cars. But Karen was completely stuck and couldn't get out! This really broke my heart. Luckily, at her primary school she's accepted completely and has lots of girlfriends. But I worry about what will happen when she goes to secondary school. I really hope a drug will become available before then so she doesn't have to struggle with obesity all her life.'

LEPTIN AS THE ULTIMATE
ANTI-OBESITY AGENT

So it turns out that fat produces the hormone leptin, and Karen's story shows the importance of a properly functioning leptin system. When leptin binds to the leptin receptor in the brain, this results in a feeling of satiety, and this keeps you from continuing to feel hungry after a meal. Other research has shown that leptin also stimulates fat burning in the body, and so leptin is also often referred to as an 'anti-obesity hormone'. But this is not the biological role of leptin; its main biological role is to tell the body (and the brain in particular) how much energy is stored in the body. So leptin functions as a kind of a 'fat sensor' inside the body!

This is a very ingenious system, because the amount of leptin produced by the fat cells is in proportion to the amount of fat stored in the fat cells. So the more fat there is, the more leptin is produced and released into the blood. A person with more body fat will then also have higher leptin levels in their blood. This fat sensor is important, because if there is little stored fat (and thus a low level of leptin in the blood), the leptin receptors in the satiety centre will receive fewer signals. And, as a result, the body will take measures to increase the amount of stored fat: you'll get hungrier!

When leptin was first discovered, researchers thought they had found the 'holy grail' of weight loss. Surely, a large proportion of people with obesity must not be able to make enough leptin, which would mean that they would often have a hard time feeling full and so also ate more than lean people? So, shortly after the discovery of leptin, leptin levels were measured in various groups of people. And guess what? The leptin levels of people with obesity are almost always very high! This is actually not surprising when you know that leptin is produced in proportion to the amount of fat a person has. So why does it often take longer for people with obesity to feel full even though they have high leptin levels? This is because of a phenomenon known as 'leptin resistance'. Leptin resistance is probably due to a combination of various factors, including inflammation, which often occurs along with obesity (more about this in Chapter 4). Consequently, leptin is less able to send its signals through the receptor, and leptin will thus be much less effective as an appetite suppressant. In other words, in spite of high leptin levels, you're still very hungry.

Then wouldn't it help to treat people with obesity with extra leptin? As long as the dosage is high enough, wouldn't it at least have some effect? In the 1990s such treatment methods were extensively tested, but unfortunately the results were disappointing. People either did not lose weight, or lost only a few kilos. Apparently, leptin resistance in obesity is much more persistent than we had hoped.

What has achieved fairly good results is administering leptin to people with obesity once they have lost a significant amount of weight. Every person has an individual 'set point' in terms of the amount of body fat the body strives to maintain. A kind of personal benchmark. If someone is below their personal benchmark, the body will activate mechanisms to increase the amount of body fat, such as slower metabolism, more hunger and especially a craving for fatty foods. This is why it's so hard to maintain your new weight once you've slimmed down, something that will be familiar to most people who've been on a strict diet. These effects occur in part because leptin levels in the blood drop once you've lost weight. And less leptin means . . . more hunger. When people with obesity who have lost more than 10 per cent of their body weight are given extra leptin, something interesting happens: now the mechanisms that ensure a return to their personal benchmark do not get activated. This means that their metabolism does not slow down, they feel full longer and actually crave fatty foods less – all thanks to an extra dash of leptin. So should everyone be taking leptin, then? Alas, we have to temper this enthusiasm, because leptin is very costly and so is only used in those rare instances where people cannot produce leptin themselves, such as the Pakistani children who benefited so greatly from this.

OTHER FAT HORMONES

The discovery of leptin was a huge breakthrough in fat research. Fat was not just a passive organ for storage. No. It was able to make one hormone, and perhaps many more, that had an effect on the rest of the body, including the brain. So, over the past twenty years, various research teams have enthusiastically devoted themselves to discovering new hormones and other substances produced by our fat. And with success! So far, more than 600 of these substances – or 'fat hormones' – have been discovered. These fat hormones have a variety of effects, although exactly what most of them do is still unclear. Some of them are inflammatory substances, others affect blood pressure, and yet others affect the body's sensitivity to insulin. Perhaps new and surprising discoveries will soon be made about the functions of some of the fat hormones.

These days, then, we like to see fat as a kind of conductor who gives various instructions (by releasing fat hormones) and in this way can exert an effect on practically all of the organs in our body. As long as the conductor does what he's supposed to do, the orchestra is in perfect harmony. But imagine what would happen if the conductor kept giving different instructions to the other musicians. The orchestra would be in complete disarray! And this is exactly what happens when our fat gets 'sick'. Because, yes, fat can get sick. This happens when the fat becomes too large, as with obesity. Then the fat produces too many of certain fat hormones, and these cause inflammation and raise the blood pressure. At the same time, the hormones that curb appetite and increase metabolism become disrupted.

One of the fat hormones discovered shortly after leptin is adiponectin, which is found at remarkably high levels in human blood. To study the function of adiponectin, mice were bred that lacked this hormone (this is a common method for figuring out what newly discovered hormones do). It turned out that mice that didn't produce adiponectin were less sensitive to insulin (you remember, the hormone that sees to it that sugars are well distributed throughout the body). As a result, they had high blood sugar, and were therefore in an early stage of diabetes (prediabetes). When the mice were given extra adiponectin, this had a number of favourable effects. The mice became more sensitive to insulin, had lower blood sugar levels and were less likely to get diabetes; they had less inflammation in their blood, and less atherosclerosis and cardiovascular disease. So adiponectin is a beneficial hormone, at least according to what has emerged from research on mice.

But listen to this: it turns out that the levels of this beneficial hormone are lower in people with obesity than they are in lean people. Furthermore, a low adiponectin level in the blood is a predictor of developing diabetes as well as of having a heart attack. Unfortunately, we don't yet know why adiponectin levels are lower in people with obesity. However, researchers do agree that increasing the amount of adiponectin in the body is beneficial. So how can we do this? One way is to lose weight. Various studies have shown that losing 10 per cent of your fat mass causes adiponectin levels in the blood to rise sharply. In addition, there are drugs that indirectly cause adiponectin to rise, including cholesterol-lowering medication and certain drugs used for diabetes. Unfortunately, administering adiponectin directly is not yet an option, seeing that adiponectin is only effective for a short period of time and that the hormone is very expensive to make. However, a great deal of effort is being put into finding drugs that can bind directly to the adiponectin receptor. This would produce the same effects as higher adiponectin levels in the blood. Although the initial results are promising, there is still a long way to go before this can be used to treat obesity.

IS FAT FERTILE?

Fat hormones have even more surprises in store. A case in point: Maggy is thirteen and an avid gymnast. She trains twenty-four hours a week, every day after school and also at the weekend. Because gymnastics is her passion, it's not hard for her to put everything else aside for this. This is in contrast to her twin sister Evelyn, who has very different hobbies – watching TV series, putting on make-up, anything that doesn't involve exercise. Although Maggy and Evelyn are exactly the same age, they are considerably different. Gymnast Maggy is more than 10 centimetres shorter than her sister, and much slimmer. Also, Maggy's breasts haven't started to develop yet and she hasn't yet started menstruating, while Evelyn had her first period just before she turned twelve. It's widely known that young top-level gymnasts are shorter and more slightly built than most girls in their age group. This could even be termed a growth delay. This delayed growth can sometimes be so pronounced that they never reach their full natural adult height. In addition, top-level gymnasts have a lower percentage of body fat as a result of a combination of low food intake (sometimes less than what they actually need) and high energy expenditure due to their intensive training. Quite surprisingly, what has also emerged is that young top-level gymnasts are older when they menstruate for the first time. While the average European girl gets her first period between the ages of 12.5 and 13.5, this is between the ages of 14.3 and 15.6 for top-level gymnasts. Why is this?

In the 1970s, this question greatly intrigued the American biologist Rose Frisch (1918–2015). Frisch discovered that girls who had little body fat, especially athletes and girls with the eating disorder anorexia, started menstruating later and were also less fertile. By carefully studying the body composition of a large group of girls, she showed that girls need a body fat percentage of at least 17 per cent to be able to start menstruating. So, the amount of body fat, not age, is the most important factor in the onset of menstruation! Moreover, this same lower threshold of 17 per cent is also necessary to keep menstruating. Adult female athletes with a lower percentage of body fat stop menstruating and as a consequence find it very difficult to get pregnant. Rose Frisch was able to tell athletes almost exactly how much weight they needed to gain in order to become fertile again. Her son Henry Frisch later said that some of these women were so grateful to his mother that they named their daughters Rose. We are still benefitting from the insights that emerged from Rose Frisch's studies. When a woman visits her gynaecologist because she can't get pregnant, her height and weight are measured if necessary to see if she might be underweight, and if she is, she is asked how much she exercises.

THE CHOCOLATE HORMONE: THE ONSET OF PUBERTY

From an evolutionary point of view, the link between little fat mass and infertility is quite logical. If you have low energy reserves, your body won't be ready to carry a child, because pregnancy consumes a lot of energy. But how does that connection between fat mass and fertility actually work? For a long time, people were in the dark about this. That is, until leptin was discovered. It turned out that the Ob mouse – which we learned about at the start of this chapter and which is unable to produce leptin – was infertile! Giving leptin to this mouse led not only to decreased appetite, but the mouse also became fertile. Suddenly, many of the puzzle pieces fell into place. Leptin, the body's fat sensor, tells the brain not only how much fat the body has stored, it is

47

also linked to the centre in the brain that influences fertility. If this centre in the brain doesn't send the right signals – and this is what happens when there is too little leptin – ovulation won't take place and so the woman will not be fertile. She will also stop menstruating.

This was also the case for Natalie from Chapter 2, who has lipodystrophy. Because she had virtually no subcutaneous fat, her body produced so little leptin that, as a result, her periods were very irregular. When things are reversed, it works the same way: if a girl has lots of fat, her body produces a large amount of leptin and she will start menstruating much sooner. Indeed, girls with overweight often start menstruating when they are still very young.

Many studies are being conducted to learn exactly how leptin is linked to the brain's 'fertility centre', and this probably takes place via substances that act as go-betweens or 'mediators' in the brain. One crucial mediator is the hormone with the romantic name kisspeptin. This hormone is responsible for the onset of puberty and you could say it 'switches on' sexual maturation. If the body has enough fat, leptin gets kisspeptin production underway, and this sets the brain's fertility centre in motion. Those people with insufficient kisspeptin production also have low levels of sex hormones in their blood and enter puberty late. And too much kisspeptin results in early onset of puberty. So, if you're a girl and a bit unlucky, your breasts might already start to develop at age eight instead of eleven! Kisspeptin is also sometimes called the 'chocolate hormone'. This is because, in the mid-1990s, this hormone was discovered by researchers in Hershey, Pennsylvania – and this is the home of Hershey's Chocolate Kisses, which are very popular in the United States.

Leptin also plays an important role during pregnancy. It so happens that the placenta also produces leptin. Furthermore, a form of leptin resistance develops during the second trimester of pregnancy in the same way as it does with obesity: in spite of high leptin levels, the leptin has less of an effect. This probably explains the eating binges that pregnant women are known to fall prey to (so it's nothing to be ashamed of!), and which are intended to build up an extra supply of fat that will be used to feed the baby once it's born. So, their mother's fat is also crucial to newborns.

It's clear. We no longer view our fat simply as an organ to be used for storage. No, our fat is smart! Helped by hundreds of different fat hormones, our fat can exert an effect on nearly every other organ in the body, including the brain. If our fat threatens to shrink in size, it uses leptin to tell the brain we should eat more. And again via the brain, it also keeps pregnancy from occurring if that would put a strain on our remaining fat reserves. These hormones make our fat into a fantastic, resourceful organ. But the 'master conductor' can sometimes also get terribly confused. When the fat becomes too large, as in the case of obesity, this causes a major disruption to hormone production in the fat, and can lead to the development of a wide range of diseases.

4

Fat can result in disease and disease can make you fat

SICK FROM TOO MUCH BODY FAT: ROB'S STORY

Rob is a sixty-five-year-old retired educational specialist. He is the father of two children, and has two grandchildren. Rob has many hobbies, and his days are filled with things like riding his motorcycle, sailing and volunteer work, activities he loves and does with great enthusiasm and energy. But things weren't always this way. 'Before I turned forty, I never had to watch what or how much I ate. I really enjoyed going out to eat, and sitting around the table after dinner with friends. I was also very satisfied with my body. But then I had a couple of setbacks. I lost my job and got a divorce, and life was hard for a while. To cope with my emotions, I started to eat more and more. During this difficult time, food gave me a sense of gratification. I gained weight gradually, until, a few years back, I weighed more than 110 kilos – at 1 metre 90 centimetres this was far too heavy for my height. I was officially "obese". And almost all of the fat had accumulated in my belly area, just like it had on my father.' Rob felt bad that he no longer had a lean body , but learned to live with it. 'That's just the way it is,' he said matter-of-factly. But he did have to regularly endure the disapproving looks of his children, who thought he should really do something about this. 'And deep inside I wanted to, but I didn't know how, and I wasn't motivated enough to take action.'

He found his motivation some time later, when he began developing symptoms that were related to his obesity. 'I had very little energy and was always so thirsty. I went to my general practitioner and it turned out I had diabetes. I took tablets to start with, but they had nasty side effects like nausea and diarrhoea, and I wanted to quit taking them. The next step was daily insulin injections.' Rob also suffered from severe sleep apnoea, and he would stop breathing more than thirty times an hour. 'As a result, I'd wake up exhausted. At night, I had to start wearing a nose mask, a device that kept my airway open and helped me breathe better. And, if that wasn't enough, it turned out I also had both high blood pressure and high cholesterol, and I had to take three different kinds of tablets for this.'

Meanwhile, Rob's blood sugar kept rising and he had to inject more and more insulin. 'I had to go to an ophthalmologist every year to see if my eyes were still okay, because you can go blind from diabetes over time. This was my biggest nightmare, because I knew someone with diabetes who was almost blind. I was worried about my blood sugar, which just kept rising, and I was referred to an internal medicine specialist.' Although he'd heard before that he needed to lose weight, he was now told that, if he lost weight, not only would he be able to reduce the amount of insulin he injected, his other problems, like sleep apnoea, high blood pressure and high cholesterol, would disappear.

Rob was finally motivated, and decided to drastically change his lifestyle. He went to a dietitian to improve his diet. He also walked his dog more often, and took the stairs instead of the lift. In addition, he received coaching for these behaviour changes. Thanks to this combined strategy, he lost more than 10 kilos. And what happened? 'I had more energy, needed much less insulin than before, and my sleep apnoea improved, which meant I woke up feeling much more refreshed. But I'm not there yet - I want to lose even more weight so

I can also stop taking my high blood pressure pills. Every day is still a struggle, though. For example, over the past few years, I've started to notice that I don't feel full as often. So I have to make a conscious effort to stop eating, because I still feel hungry – after finishing a whole plate of food, I could easily eat another plateful.' Although he still has a long way to go, he keeps moving towards his goal through all the ups and downs!

THE LIFE CYCLE OF FAT

Rob's excess body fat literally made him sick. And he is by no means alone. It is estimated that roughly half of all people with overweight have one or more complications related to this, such as high blood pressure, sleep apnoea, high cholesterol, or diabetes. And, over time, a vast majority of all people who are severely overweight (obese) will develop these conditions. To understand exactly how this works, it's helpful to have a closer look at what our fat actually consists of. As we know by now, it is made up of fat cells, which, as we saw in Chapter 2, can best be compared to tiny balloons that become filled with fat. And there are lots of them: a whopping 50 billion! But that's not all. Our fat also contains inflammatory cells (white blood cells), which are found in between the fat cells. Like Pac-Man, these cells are on standby, ready to eat up pathogens like bacteria and put them out of commission. They also gobble up dead or malfunctioning fat cells to keep the fat organ from becoming a shambles and to make sure it continues to function properly. And if we're facing a serious threat, they can also sound the alarm so that auxiliary troops (even more inflammatory cells) can come into action. This alarm is triggered by inflammatory substances, which can be released by the inflammatory cells both locally and in the blood. Inflammatory cells are vitally important. If you didn't have inflammatory cells, the tiniest intruder – whether in the form of a virus, bacteria or fungus – would be deadly. This is why, if left untreated, the disease AIDS – in which the HIV virus deactivates a portion of the inflammatory cells – is always fatal in the end. But no matter how important inflammatory cells may be, they also have a dark side – for example, when large numbers of them accumulate in our fat. When this happens, they continue to release so many inflammatory substances that large quantities of them end up in the blood, and this can have widespread effects. We call this 'subclinical inflammation' because, although there is no real inflammation, such as we'd have for a festering wound or injury, inflammatory signals are constantly being released into the blood. These inflammatory substances are one of the culprits in the link between obesity and various diseases. They're harmful not only to your cardiovascular system, they can also find their way into your brain and result in low mood, or even depression. We'll discuss this in more detail later on.

Along with inflammatory cells, our fat tissue also contains cells that make up the walls of tiny blood vessels. These minuscule blood vessels transport glucose, fatty acids and also oxygen to the fat tissue. Fat also contains nerve endings. These thin strands connect our brain to our fat, and our brain can use them to send instructions to our fat – for example, when fats need to be released into the blood and sent to other organs. And finally there are stem cells, 'primal' cells that have the ability to develop into a variety of different cell types.

When someone with a slim build becomes overweight or obese over time, their fat undergoes a fascinating transformation, which begins with the fat cells themselves. Imagine that we were to take a 'fat biopsy' (remove a piece of fat) from the belly of a slender person and also take one from the belly of an overweight person. If we then examined these pieces of fat under the microscope, we would notice a number of important differences. First of all – and this will probably no longer come as a surprise – the fat cells from the overweight person will be larger, since more fat is being stored in them. Also, a person with overweight often (but not always) has more fat cells.

For a long time we thought that, over the course of a lifetime, both the size and the number of fat cells could vary. In other words, if the amount of fat in your body were to increase over a period of time, your fat cells would get larger and also the number of fat cells would increase. And if you were then to lose weight, both the size and the number of fat cells would decrease. Although this all sounds terribly logical, it turned out that this hypothesis was inaccurate! In the 1970s, a remarkable experiment was carried out to test this hypothesis. A number of lean young men between the ages of twenty and thirty took part in a study in which, over the course of four months, they first had to gain weight by eating more and exercising less. During this period, they gained an average of 10 kilos of fat. Then they had to shed those kilos by eating less and exercising more. While they were losing weight, biopsies of their subcutaneous fat were taken from various locations at a number of different moments. The results were staggering. When the men gained weight, only the size of the fat cells increased – the number of fat cells remained exactly the same. When they subsequently lost the weight, the size of the fat cells decreased and the number of fat cells again stayed the same. Later studies confirmed these results. In people who lost a great deal of fat mass as the result of bariatric surgery, two years after the operation only the size of the fat cells had decreased, while the number of fat cells remained the same. It thus seems that, once the number of fat cells in our body has been determined, it's impossible for us to get rid of them!

So when is this number of fat cells determined? A research team from Sweden tackled this question. They took fat biopsies from people of different ages (between zero and sixty years of age) with varying amounts of fat mass, and looked at the number of fat cells these pieces contained. So, listen to this. They discovered that the number of fat cells increases during both childhood and puberty. From around the age of twenty, this levels off and the number of fat cells remains constant. Apparently, childhood is the most important period when it comes to determining – or 'programming' – the number of fat cells. The number of fat cells will increase more quickly in a child who is overweight than in a child who is not. And if the child remains overweight, the child will have a larger number of fat cells for the rest of their life and will never be able to lose them. Our body is amazingly good at keeping the number of fat cells constant. Even after liposuction – in which fat cells are literally sucked from the body – the body performs a cunning trick, and the fat cells return in another location! Numerous studies in women have shown that this fat often turns up in their breasts. Of the women who have had liposuction, 40 per cent go up one or more cup sizes. Although this will be welcomed by some, this phenomenon is nothing short of remarkable. So there must be some kind of system that records the number of fat cells in our body and takes action when this number drops. This is something that is actually happening all the time, because, just like many other types of cells in our body, fat cells die and are then replaced by new ones generated from stem cells (the primal cells in our body fat mentioned earlier). It is estimated that, every year, 10 per cent of the total number of fat cells in our body are replaced by new ones. In other words, after ten years, your fat tissue will have been completely renewed. However, the body makes sure that exactly the right number of new fat cells are created so that its total number will remain constant. Precisely how this happens is still a mystery. But apparently it was important enough from an evolutionary point of view to invest in this – as long as we can store enough fat!

So how bad is it to leave childhood and enter adulthood with a greater number of fat cells? Unfortunately, this makes it more difficult to slim down and to stay slim as an adult. Because, if you have more fat cells and then lose weight, your fat tissue will contain a large number of small fat cells. We have already seen that the larger a fat cell is, the more of the fat hormone leptin it releases. A large number of small fat cells will produce and release less leptin in total. Since leptin curbs appetite, this will then lead to greater appetite and decreased fat burning – anything to return to the 'old' situation in which the fat cells were fuller. Unfortunately, this is often what happens. This is one of the main reasons why it's so difficult to become a slender adult if you have been overweight as a child.

PEOPLE ACCUMULATE FAT DIFFERENTLY: APPLES AND PEARS

We often view 'body fat' as one big organ, an enormous mass of fat cells that stores fat and produces hormones. But this isn't entirely accurate. Fat actually does other things, depending on where it is located. As mentioned earlier, the two largest quantities of fat are found in the belly around the organs ('belly fat') and just beneath the skin ('subcutaneous fat'). You've probably noticed that people differ considerably when it comes to where they accumulate the most fat. There are even special names for this. If you have lots of belly fat, you are said to be 'apple-shaped', which is common in men like Rob. If a person instead has lots of subcutaneous fat around the buttocks, hips and thighs, they are said to be 'pear-shaped', something we tend to see more in women. Fat accumulates in smaller quantities in many other places, such as around the heart, kidneys and even the blood vessels.

These differences in body fat distribution (apple- versus pear-shaped) were first described in 1956 by Jean Vague, a doctor in Marseille in France. He had also noticed that people with more belly fat had a higher risk of diabetes than people who instead carried more fat around their hips and thighs. What's more, a number of later studies appear to indicate that the presence of lots of fat around the hips actually protects against diabetes! So, belly fat is worse than (subcutaneous) hip fat. But why?

WHY BELLY FAT IS WORSE THAN HIP FAT

The fat cells in belly fat are tucked in between the organs. As a result, they are less able to stretch, and so also store less fat. But . . . that's a good thing, right? Afraid not. When fat cells have reached what is known as their 'maximum storage capacity', the excess fatty acids in the blood will have to be stored elsewhere (see **Box 6**). The body doesn't have an advanced system that would automatically channel this to the subcutaneous fat. This would be ideal. Instead, it's transported to other locations, such as the muscles, liver and around the heart. Extra fat is not welcome here, to say the least. It actually doesn't belong here at all. Fat deposits will interfere with the functioning of every one of these organs. It will disrupt sugar metabolism in muscle and in the liver, which means that they will become less sensitive to the effects of insulin, the hormone that makes sure the 'sugar gates' open so that the cells can absorb sugars. We call this insensitivity to insulin 'insulin resistance'. As a result, the organs absorb less glucose from the blood, and the blood sugar levels rise, and even more insulin is released to deal with the high blood sugar levels. If this goes on for too long and the pancreas is no longer able to cope by releasing extra insulin, this will ultimately lead to a situation in which the blood sugar levels rise to dangerously high levels, and diabetes will result. Extra fat accumulation around the heart can also interfere with the heart's pumping function.

Box 6. The 'elastic fat cell' theory

Not all fat cells are the same. Some people have very 'flexible' and 'elastic' fat cells. Others, however, have fat cells that are more stiff, and so they fill up faster. Think of them as balloons

---

made by different companies, where one balloon can be inflated larger than the other. This means that, in those whose fat cells are more stiff, the fat will be more likely to overflow to other organs. These people are therefore also more likely to develop complications related to their obesity, such as insulin resistance and diabetes. We call this the 'elastic fat cell' theory, and it's likely influenced by genetics.

Although an increase in belly fat produces the most harmful effects, when fat expands in other places, it can also cause problems. Because, whenever there is an expansion in fat tissue, whether this is belly fat or subcutaneous fat, this always disrupts the structure of the fat tissue, which is formed from structural proteins. There is also an entire network of blood vessels between the fat cells. When the fat cells become filled with fat, the fat will expand and cause this network of blood vessels to stretch. To be able to continue to supply all of the fat cells with blood, and thus oxygen, extra blood vessels need to be formed. If not enough blood vessels are formed or if this takes too long, the fat cells in the middle won't get enough oxygen. This shortage of oxygen is very harmful for fat cells, and cells will die off. And this is where the inflammatory cells – the little Pac-Men inside our body – come in! The presence of dead fat cells attracts inflammatory cells in the same way that maggots are attracted to dead or rotting meat.

What stands out is that, when we compare a piece of fat from a slender person with that of a person with obesity, the person with obesity's fat often contains many more inflammatory cells (see **Figure 4**).

The inflammatory substances released by these cells attract even more inflammatory cells: Come on, guys! Let's get to work here! And this process can take place in both subcutaneous fat and belly fat. The fat cells themselves can also release inflammatory substances in response.

However, one significant difference here is that fat cells within belly fat will release more inflammatory substances than those within subcutaneous fat. This is yet another reason why belly fat is more harmful than subcutaneous fat. This explosion of inflammatory substances can have many consequences. They are released into the blood and can lead to insulin resistance in organs like the muscles and the liver, and thus result in elevated blood sugar levels. Moreover, inflammatory substances in the vicinity of blood vessels can also lead to inflammation of the vascular wall, and this can ultimately result in cardiovascular disease.

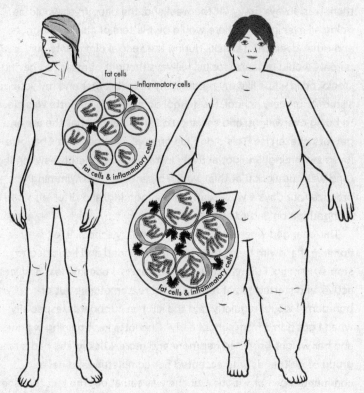

Figure 4. Inflammatory cells in the fat of a slender person and a person with obesity.

BEING OVERWEIGHT MAKES YOU LESS FERTILE: CHARLOTTE'S STORY

When body fat increases, this also affects fertility. Charlotte, a cheerful thirty-three-year-old woman, knows all about this.

'I come from a family that loves to eat, and food has always played a central role in our life. We always took the time to enjoy our food, and taking second helpings was considered completely normal. My mother would always fry the meat in a generous amount of fat, and there was always gravy. At the weekend, the deep fryer would be going all evening, and there would be platters of chicken nuggets and other deep-fried snacks during the special film nights our parents would organize for us. I always thought of myself as having a stocky build – just like my parents, my older brother and my younger sister.' At primary school, the school doctor said Charlotte was close to being overweight, and she should try to watch what she ate and get lots of exercise. This didn't fall on deaf ears, because Charlotte loved swimming! She got actively involved in competitive swimming (and was quite good at this), which meant she had swimming practice four days a week, early in the morning, and often also had a competition on Saturday.

This changed when she went to secondary school. 'I felt very at home in that swim team, but sadly it disbanded and I decided to stop swimming. I didn't like any other sports. I became less and less active, and started to eat more and more, sometimes out of boredom. I would regularly raid the kitchen cupboards, especially when I got home from school early.' Charlotte kept gaining weight, and her weight bothered her more and more. Luckily, she had a nice group of girlfriends who accepted her completely, and never commented on her weight. But she was sad about the fact that she had never had a boyfriend. Charlotte went off to study law, and in one of her classes she met Arthur, the love of her life. 'Arthur was also "plus-sized" and he thought I was gorgeous just as I was. After a couple of years, we decided to live together, and moved into a pleasant apartment in the middle of town. We would regularly invite our friends over for dinner, and we would get through many calories in the form of bottles of wine and lots of food. We also liked to make things cosy during our evenings on the sofa at home, and French cheeses were a particular favourite.'

As their love grew, so did their girth, until both Charlotte and Arthur had obesity. In the meantime, they had graduated and had both found good jobs – Charlotte as a legal advisor for a large company, and Arthur with the government. 'And then . . . we wanted to have a baby. I had always had a hormone IUD, and, with the exception of some spotting, only rarely got my period. One rainy November day, I had my IUD removed, an almost legendary moment for us. From now on, it could happen at any time. We were ready! I waited patiently for my first menstruation. One month, then two months passed. But I didn't get my period. Had we succeeded "the first time round"? Was it possible I was already pregnant? But no – the pregnancy test was negative. Four months later, I finally got my first period. But I didn't get it again for three months. And the pregnancy test continued to indicate that I wasn't pregnant. I started to get worried! After more than a year, we went to our general practitioner, who told me that my being overweight probably had something to do with it, and advised me to lose weight, ideally until I had a healthy BMI. Well, that was easier said than done. We went home very disappointed – a baby seemed further away than ever.'

Being overweight is known to lower the chance of becoming pregnant. This is mainly because a woman who is overweight has more irregular periods because she ovulates less often. Or her ovulation may stop entirely, which means she won't be able to get pregnant at all, at least temporarily. This is a direct result of the increase in the amount of body fat. When body fat increases, the conductor gets completely confused – the production of fat hormones gets out of balance, including the production of the fat hormone leptin. We saw earlier that when not enough leptin is produced, as with underweight, this leads to infertility because the brain's 'fertility centre' can no longer be switched on. But the opposite also turns out to be true. Too much leptin – which is nearly always the case with overweight – can lead to infertility as well, because this also disrupts the functioning of the fertility centre. So, both too much and too little leptin can reduce the frequency of ovulation or stop it entirely. There are also many other fat hormones that affect fertility, and more or less the same principle applies to each of them: an imbalance will inhibit fertility in women.

With overweight and obesity, something else happens as well, something that is of even greater importance. When body fat expands, the activity of the enzyme aromatase also increases in the fat cells. In both men and women, this enzyme converts androgens (male hormones such as testosterone) into female hormones known as oestrogens. Oestrogen has many important roles in the body, one being thickening of the uterine lining (getting it ready in case of pregnancy), but too much oestrogen can inhibit the fertility centre in the brain. In other words, when the amount of oestrogen in the body gets too high, this suppresses ovulation. This is one of the ways in which birth-control pills that contain oestrogens can prevent pregnancy. So, too much fat mass in fact works in the same way as birth-control pills.

Overweight and obesity reduce fertility not only in women, but also in men. Just as in women, when men's body fat increases, the protein aromatase becomes more active. Again, this leads to increased production of oestrogens and an imbalance between male and female hormones (often with not enough testosterone). To put it in simple terms, a man becomes 'feminized'. This results in decreased sperm cell production, and thus reduced fertility. Nowadays, men also become less fertile because they sit a lot and wear tight trousers or skinny jeans. As a result, the temperature in the scrotum, which contains the testicles, gets too high. A man produces the best 'swimmers' when the temperature in the scrotum is between 34°C and 35°C. So, if you'd like to improve your fertility as a man, try to stand more (that's good for a number of reasons – see Chapter 6) and throw away those tight trousers. And the same thing actually happens with overweight. Lots of fat in the lower belly and thighs will also lead to a higher temperature in the scrotum. Along with various hormonal changes, this is one of the reasons why increased fat in men leads to fewer sperm cells, which are also of lower quality.

Imagine that, like Charlotte and Arthur, you find yourself in a situation in which overweight is interfering with fertility. Is there something you can do about this? Absolutely! Studies have shown that weight loss – by eating a healthy diet, exercising more or, even better, a combination of the two – has beneficial effects on fertility, in both men and women. Ovulation returns, menstrual periods become more regular and the likelihood of getting pregnant increases. Even losing 5 to 10 per cent of your body weight has positive effects.

This was also what Charlotte's general practitioner told her. Try to lose weight! That was incredibly difficult, because Charlotte had been too heavy her whole life. But when more and more of the women around her were getting pregnant – colleagues, women friends, even her sister-in-law – she'd had enough. The fact that others were getting pregnant and she wasn't affected her so much that she came up with a plan to lose 25 kilos. In consultation with a dietitian, she put together a diet and went to the gym three times a week, where she exercised under supervision. In eight months' time, she had lost 25 kilos. And what happened? Her menstrual cycle became regular again. Out of solidarity, Arthur also decided to watch what he ate and occasionally joined her when she went to the gym.

He lost 10 kilos as a result. And a few months later, the pregnancy test finally showed 'pregnant'! Charlotte and Arthur are now the proud parents of two sons. And those kilos? Some of them have returned, but they've never been happier.

OVERWEIGHT INCREASES
THE RISK OF CANCER

We know Rose Frisch as the tireless researcher who showed that athletes stop menstruating if they have a body fat percentage lower than around 17 per cent. But she made even more ground-breaking discoveries, one of which again concerned top female athletes. Frisch showed that women who had been top athletes had a lower risk of developing breast cancer and cancer of the reproductive organs (such as uterine cancer) during their lifetime. And other researchers have shown that overweight and obesity increase the risk of cancer. So far, we know of thirteen types of cancer that those who are overweight or have obesity are more at risk of developing. These include uterine cancer, breast cancer (post-menopause) and ovarian cancer in women, and prostate cancer in men. Of these thirteen types of cancer, 20 per cent of the cases in Europe – so, one in five – are thought to actually be caused by overweight and obesity. Why is this?

One of the causes can be found in the greater quantities of inflammatory substances that are released as the body fat expands, because these substances can promote the growth of cancer cells. It's interesting to note that anti-inflammatory medication, such as the painkiller aspirin, can help to prevent the occurrence of certain types of cancers (such as colon cancer). As explained earlier, when fat expands it also produces more oestrogen. In women, high lifetime exposure to oestrogens increases the risk of breast cancer and uterine cancer. Seeing that top female athletes with little body fat produce less oestrogen, this explains their reduced risk of such things as breast cancer, as Rose Frisch showed.

There is also good news. A large Swedish study in which more than 4,000 patients with obesity participated showed that reducing the amount of body fat can also reduce your risk of cancer. Roughly half of these patients underwent bariatric surgery (such as gastric bypass surgery) and the other half did not. These patients were followed for eleven years. During this period, those in the group that had had bariatric surgery lost 20 kilos on average, while those in the control group gained an average of 1.5 kilos. And it turned out that the cancer risk among the patients in the bariatric surgery group had dropped by 40 per cent!

THYROID HORMONE:
A CATALYST OF METABOLISM

Not only can body fat literally make you sick, the opposite is also true. Various diseases and disruptions in the hormonal system can have an effect on our fat. One familiar example is disruption of the thyroid hormone. Nearly every cell in our body needs thyroid hormone to be able to function properly. Think of it as the hormone that keeps everything moving. If you have a thyroid hormone shortage, many organs don't work properly in that they function more slowly than they should. Your heart beats more slowly, your gut digests food more slowly, which makes you constipated, and your brown fat (a kind of fat that burns fat to produce heat – see Chapter 6) doesn't function as well. Partly because of this, your metabolism slows down and you get cold quickly, even though your surroundings might be comfortably warm. Weight gain is often the first thing people notice when they have an underactive thyroid. This makes sense: your metabolism is slower, so if you keep eating the same amount, you'll gain weight. An underactive thyroid is relatively common, and currently millions of people in Europe and approximately 12 million people in the US have this condition. This is why people who gain weight for no apparent reason (for example, the person hasn't started to eat more than they used to) undergo a test to determine the level of thyroid hormone in their blood. Treatment for an underactive thyroid consists of tablets that contain extra thyroid hormone.

Your thyroid can also be overactive. As a result, there is too much thyroid hormone in the blood, and all of the organs go into fifth gear. Your heart starts to beat too quickly (which is sometimes experienced as heart palpitations), you get diarrhoea, you get hot and you feel agitated. Because the metabolism is supercharged, people lose weight, sometimes as much as 10 kilos in two months. Even so, people often still have a big appetite. The initial treatment for an overactive thyroid is a drug that inhibits the production of thyroid hormone. Changes in the thyroid hormone levels in our blood can have major effects on our body fat. However, if you do have obesity, don't be tempted to take extra thyroid hormone if your thyroid is functioning normally. This has not been shown to be effective for losing weight, and can even have nasty side effects, including for the heart and bones.

SEX HORMONES REGULATE WHERE OUR FAT ACCUMULATES

As we have already seen, there is a clear difference between men and women with regard to how our body fat is distributed: men (especially older men) carry more fat in their belly, and women carry more beneath the skin around the hips and buttocks. This difference develops during puberty, particularly in women, who from that point on will accumulate more fat on their hips and buttocks. Later in life, men develop more fat in their belly. This difference is probably caused by the sex hormones – oestrogen in women, and testosterone in men – and particularly as a result of the balance between the two. From the time a girl enters puberty, her body makes more oestrogens, and as a result more fat cells are formed around the hips and buttocks, at least until she's around the age of twenty. Women also gain weight during pregnancy. These kilos consist not only of the extra fluid and tissue that make up the growing foetus, but also additional fat mass. This extra fat will be used as a reserve supply after delivery, when energy has to be available for breastfeeding. Some women gain much more weight than others during pregnancy, and unfortunately around 10 to 15 per cent of these women retain this weight after delivery.

For many women, menopause is marked by hot flashes, mood swings and . . . weight gain. This weight gain may well be what many women find the hardest. But why does this happen? When a woman enters menopause, her oestrogen level falls, and this drop in oestrogen slows metabolism in the body. So, when a woman continues to eat the same amount of food, she will inadvertently gain weight. We also often see that menopausal women accumulate more belly fat in particular, and often just can't seem to get rid of it. So what's an effective solution to this? More exercise! The fact that fat likes to accumulate in the belly after menopause is probably caused by changes in the ratio of oestrogen to testosterone, in which the male hormone testosterone gets the upper hand. And, in women, testosterone is responsible for greater fat accumulation in the belly. This is also demonstrated by the fact that women who make too much testosterone, as in polycystic ovarian syndrome (PCOS), also develop a body shape with more belly fat and narrow hips. In men, however, a high testosterone level results in more muscle mass and less belly fat. In short, when it comes to a healthy distribution of fat, high testosterone is beneficial for men but detrimental for women. In addition, when oestrogen levels in women drop after menopause, the risk of cardiovascular disease rises. This is likely due to a combination of a rise in cholesterol levels and blood pressure, which is mainly due to the changes in hormone balance, slower metabolism and subsequent altered fat distribution.

As men age, their blood testosterone levels drop. This causes problems not only with sexual potency, but also results in loss of muscle mass and gaining a belly. In any case, there is an interesting association between low testosterone levels in men and overweight. We know that severe overweight can lead to a low testosterone level, which has to do with the conversion of androgens to oestrogen we mentioned earlier. However, the opposite also occurs: if testosterone is extremely low – for example, because the testicles (where testosterone is made) are not functioning properly – men can become overweight more easily. If a severe testosterone deficiency has developed for such a reason, testosterone replacement therapy can help men lose fat mass. However, the low testosterone is usually caused by the overweight itself. That said, this doesn't mean that all overweight men should get extra testosterone (for example, in the form of testosterone gel) – only men with extremely low testosterone levels benefit from this.

You now know that when our body fat increases, this is anything but harmless: a range of diseases from diabetes to cancer can develop via a complex mechanism involving such things as a disruption in the balance of fat hormones and the release of inflammatory substances. The amount of fat stored in our body fat is regulated by the balance between two main determinants: the amount of energy that enters our body (which in turn is driven by our appetite) and our body's metabolism. We have already seen that leptin – the first fat hormone to be discovered – has a major effect on appetite. But appetite is much more complicated, and has plenty of surprises in store. You can read all about this in the next chapter.

5

How do feelings of hunger and satiety work?

RUNAWAY EATING: JACK'S STORY

Jack is twenty-two years old and weighs 140 kilos. He's been on a diet his whole life, to no avail – he's still battling the kilos. His mother Mary recounted the following: 'From the time Jack was a baby, I had the feeling that something was wrong with him. When I breastfed him, he just wouldn't let go. Instead of the usual twenty minutes, feeding him often took me forty-five minutes. I'd never experienced that with my other children. Even then, he would cry when he had to stop! He would often cry until it was time for the next feeding. Nothing helped, not even when I switched to infant formula for hungry babies. At twelve months, he was a big baby, but not terribly fat. In the hospital they called him a "giant baby". But that's when things started to go wrong. Even though we fed him normal portions, he started to gain a lot of weight – sometimes as much as 3 kilos a month! When he was a year and a half, he weighed 18 kilos, and at two and a half he was up to 32 kilos. I was desperate.'

Although Mary had the feeling there was something physically wrong with Jack, they were first sent to a psychologist to see whether it had anything to do with her and her husband. Video recordings were made in their home to see how the family ate and what the atmosphere was like. To Mary's great relief, everything was found to be completely normal. However, this still didn't solve the problem, and Jack remained fixated on food. Mary: 'One time, I'd bought a fresh loaf of bread and left it to cool on the dining table, with the bag open. When I wasn't looking, the bread suddenly disappeared. And then I saw Jack. He was sitting in front of the television finishing off the entire loaf of bread. This couldn't be good!'

But things got even worse. When Jack was ten years old, the school called his parents to tell them that Jack had stolen money. To buy food! This was naturally a huge shock. Jack was admitted to a diagnostic centre, where he was diagnosed with PDD-NOS (Pervasive Developmental Disorder – Not Otherwise Specified), a type of autism Jack had been diagnosed with in the past. But this diagnosis could not explain his fixation on food. Even after spending months in the diagnostic centre, no one could tell Jack's mother what was wrong with her son. Then a boy was admitted who, like Jack, was interested only in eating. This boy was known to have a genetic abnormality that kept him from feeling full, and only then did a light go on. Was it possible that Jack had the same problem?

His parents took Jack, who was by now twelve years old, to paediatric endocrinologist Erica van den Akker at Erasmus MC-Sophia Children's Hospital in the Dutch city of Rotterdam, where his blood was tested. Six months later, the results were in. It turned out that Jack did indeed have a genetic abnormality, which meant that he was missing what is known as the melanocortin 4 (MC4) receptor, a receptor in the brain responsible for curbing feelings of hunger. Because this did not happen in his case, Jack was always hungry. Moreover, Jack's metabolism also turned out to be slow.

He would have to be on a low-calorie diet for the rest of his life. But even though he was only allowed to eat 1,150 calories a day – around half of what other children his age ate – he kept gaining weight. His mother recalls how terribly frustrating this was, for them as parents, but especially for Jack. It was a difficult time. Jack was often so hungry, he would be crafty when it came to food, and came up with clever ways of getting more. For example, one time he collected empty bottles and returned them for the deposit, which he then used to buy bags of sweets. Or he would get people to buy food for him.

Mary: 'Luckily, he was never really bullied about his weight, although I do sometimes notice the looks people give him. In the meantime, we've been able to find a good balance. His weight has been stable for a couple of years now, but he's still very overweight. He doesn't talk about it much, probably because of his autistic tendencies.'

Jack himself is reasonably satisfied with the balance he has achieved. Although he'd naturally like to be somewhat thinner, he can live with the way things are now - even though he's had diabetes since he was nineteen, and has to take medication for this. What he finds particularly difficult is that he's always thinking about food, especially when he's bored. Jack: 'If a pill would come onto the market that would cure my genetic disorder, I'd take it immediately! But I look forward to the things that are still ahead of me. I'd like to become an architect, and maybe design my own house.'

APPETITE: A COMPLEX INTERACTION BETWEEN HORMONES AND CONNECTIONS

You will have noticed that Jack's story is very similar to Karen's. Their stories clearly illustrate that feelings of hunger and satiety are regulated to a great degree by an interaction between appetite hormones and connections in the brain. Both suffer from a monogenic type of obesity. Leptin – which was the first fat hormone to be discovered – is one of the most important appetite hormones, and, as such, forms a link within a larger system. Leptin sends a signal to the brain that curbs appetite and provides the brain with information about the amount of fat in the body. It is a crucial hormone that has an influence on our nutritional status.

Let's have a closer look at this comprehensive system for regulating appetite and satiety. Not only does our fat send signals to the brain about our nutritional status, the gut also sends signals to the brain through the nerves that influence appetite. The main control centre for appetite and satiety in the brain is the area known as the hypothalamus, which we mentioned earlier. The hypothalamus is located deep within the brain at approximately the level of the nose, and is a real jack-of-all-trades. You can compare the hypothalamus to air traffic control at a major airport: all kinds of information is received simultaneously, and this has to be translated into a wide variety of actions in a flash. The control tower monitors the planes that are taking off and landing, and also the other vehicles on the tarmac, such as the baggage carts and mobile aircraft stairs, and even the weather conditions. In a similar way, the hypothalamus receives a huge amount of information that has to be handled properly.

The hypothalamus gets information about food intake over the long term (how much stored fat do I have?) and also about food intake in the short term (what did I just put in my mouth and what should I do with it?). Furthermore, this 'air traffic control system' also receives signals from all over the body – including from the gastrointestinal tract, the pancreas and fat tissue – that provide feedback about how much energy is being used. All chemical messengers that enter the hypothalamus have their own 'gate', known as a nucleus. The fat cells use leptin to tell the control centre, 'There are more than enough fat reserves, so turn down the appetite a little and turn up metabolism.' And the 'hunger hormone' ghrelin comes flying in from the stomach with the message, 'The stomach has been empty for quite a while now, so go ahead and switch on feelings of hunger' (more on this further along, in **Figure 5**).

What this comes down to is that a multitude of signals are constantly arriving in the hypothalamus, and the net effect of all of this is that feelings of hunger are sometimes curbed and at other times stimulated, and that metabolism is turned down or up. The combination of these effects leads to fat either being stored or burned.

If we take a closer look at our brain's control tower and all of its 'gates', it becomes clear how the feeling of satiety, or fullness, comes about. A certain region in our hypothalamus contains brain cells that make various substances that either curb or stimulate appetite, depending on the route taken. In this area of the brain, the leptin that is produced in the fat is received through its own 'gate': the leptin receptor. There, a 'cascade' of substances is set in motion, like falling dominos, with one substance stimulating the production of the next: a process that ultimately results in the feeling of satiety.

Things can go wrong at various points along the row of falling dominos, as we saw in the stories of Karen and Jack. In Karen's case, leptin is unable to bind with the leptin receptor. In Jack's case, things go wrong a few dominos further along. Usually, one of the substances released by the hypothalamus would bind to the MC4 receptor in another part of the hypothalamus (see **Figure 5**) to produce a feeling of satiety. But because Jack has a genetic abnormality that means he is missing this MC4 receptor, he can't stop eating. What it comes down to is that, with Jack, things go wrong at a different level in the hypothalamus than they do with Karen. In the past, we assumed these genetic mutations were rare, and not many people know that they exist. However, we recently analysed the DNA of 1,230 Dutch persons with obesity, and found that at least 4 per cent of them had specific gene mutations that cause obesity. Worldwide, many individuals with obesity may be walking around not knowing that they also have such a monogenic form of obesity. In the future, we hope that knowledge about the existence of these diseases lead to better treatment strategies. Diagnosis is not only important to reduce the obesity stigma these people often suffer from (more about this in Chapter 11), but also treatment can be individualized, as new drug therapies are currently available for some of them and new medications are being developed.

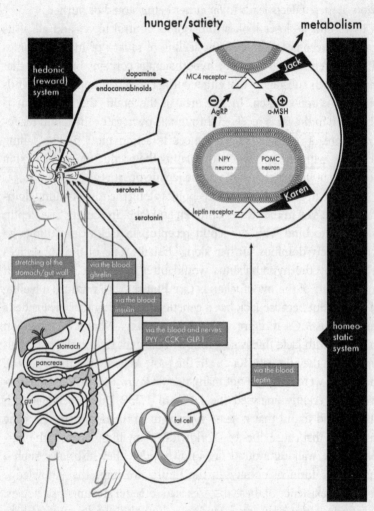

Figure 5. The appetite regulation system in humans: how are our feelings of hunger, feelings of satiety and our metabolism regulated?

When we eat, the gastrointestinal tract sends signals to our brain via hormones (like the hunger hormone ghrelin and the satiety hormones peptide YY (PYY), cholecystokinin (CCK), and glucagon-like peptide 1 (GLP-1)) and also through special receptors that measure how far the gut is being stretched ('mechanoreceptors'). At the same time, the pancreas sends signals via the hormones glucagon and insulin, among others. And fat tissue also chimes in via hormones like leptin and adiponectin. Although all of these chemical messengers arrive at the same busy airport in the hypothalamus, they each arrive at their own gate, or 'nucleus'. And they each have their own message to communicate, and help to determine whether the appetite has to be curbed or stimulated at that moment, and whether it's time to turn metabolism up a little or turn it down. This is what is known as the homeostatic system. If these satiety hormones have already caused you to feel full and then a luscious dessert appears, thanks to your hedonic system (otherwise known as your reward system), you'll still be able to eat that dessert! In both Karen and Jack, important receptors for these satiety hormones are defective, which means they're always very hungry.

FOOD CHOICES ARE OFTEN UNCONSCIOUS

Did you have any idea that, while you were innocently munching your sandwich at lunch, this kind of domino effect had been set in motion that would ultimately ensure that you would just eat one or two sandwiches and not the entire loaf? So it's not just a question of how full your stomach is that prompts you to stop eating, as many people believe. Feelings of hunger and satiety are the result of an ingenious interaction between hormones and nerve cells. And the brain takes the final decision.

Research has shown that every day we make on average around 220 food choices, and most of them are unconscious. These choices are driven in part by exposure to the food in your immediate environment and the way in which your body responds to this. If you see a delicious doughnut or smell a freshly baked cake – or even just think about one – your mouth might start to water. Say that you're crazy about chocolate – in that case, just the thought of a chocolate bar can wake up the hormones in your body and cause the insulin level in your blood to rise sharply, and, as a result, your blood sugar level will drop and you'll actually find yourself in physical need of sugar. And if you then arrive home after a long day at work and you have to choose between a bar of chocolate or a salad, your body's response will be more likely to steer you in the direction of the chocolate bar. Even though you know very well that the salad is much healthier.

The signal to stop eating also happens unconsciously to a large extent, and is strongly influenced by our hormones. As mentioned earlier, we don't stop eating simply because our stomach is full and it's physically impossible to fit in any more food. We stop eating mainly because food enters our small intestine, and then signals are sent to the hypothalamus via nerves and hormones. Only once these signals have been clearly received does a feeling of satiety arise and we stop eating. Some people feel full much faster than others, and in rare cases (such as with Karen and Jack) this doesn't happen at all. These differences are determined in part by our genes. Just as genetic variation determines whether your hair will be curly or straight, and whether your eyes will be blue, brown or green, one person will tend to feel full quickly, while for another person this will take longer. For the person who takes longer to feel full, there is a greater chance that they will take second and maybe even third helpings at mealtimes, and they are also more likely to be heavier.

Just imagine how hard it must be to never feel full! Always walking around feeling hungry in a world in which food is there for the taking, everywhere you look. And much of this food is also high in sugar and fat! Karen's paediatric endocrinologist Erica van den Akker compares the experience of constantly feeling hungry to walking through a blazing desert. 'You're completely dehydrated and dying of thirst. Your mouth is so dry that your tongue sticks to the roof of your mouth. Then suddenly you see a delicious glass of cool water – but you're not allowed to touch it.'

THE STOMACH'S HUNGER HORMONE: GHRELIN

In the previous section, we talked mainly about our feeling of satiety. But what triggers our feeling of hunger? If you've been walking around with an empty stomach for a while, at a certain point, it's time to eat again. You will feel the urge to eat something when ghrelin – the stomach's hunger hormone – comes into action and sends a signal to the brain. Ghrelin is thus the opposite of leptin, which we consider to be the satiety hormone. Think of that freshly baked cake we mentioned earlier and how hungry this can make you . . . Ghrelin amplifies this phenomenon.

It's interesting to note that ghrelin production peaks right before a meal, and drops sharply within an hour of eating. When ghrelin concentration rises, this leads to the production of two signalling proteins ('neurotransmitters') in the hypothalamus: neuropeptide Y (NPY) and agouti-related peptide (AgRP). These trigger feelings of hunger in the brain. What is unique about ghrelin is that this is the only hunger hormone that the body sends to the brain. All of the other chemical messengers that trigger hunger are made in the brain itself. This is different for the satiety hormones, which are produced in a number of places, including the brain, the gut and fat tissue.

Ghrelin also stimulates the production of gastric acid, the movements of the gastrointestinal tract, and the emptying of the stomach. This is how our gastrointestinal tract prepares for the food that will need to be digested. If you have an empty stomach, you can sometimes already hear the movements in your stomach or gut. When your stomach growls, gastric juices are on the move (and an empty stomach in particular functions in a way similar to the soundbox of a guitar). A growling stomach isn't necessarily a sign of hunger, and so you don't need to eat something right away. It could be that your body is used to being fed at a certain time, and is already getting ready for this. Habits can have a strong effect on our body!

It seems almost magical that our thoughts can influence our ghrelin levels. Psychologists at Yale University conducted a fascinating experiment. They gave forty-six research participants a milkshake that contained 380 kilocalories. Half of the participants were told that it was a very high-calorie milkshake, with 620 kilocalories, and the other half were told that it was a 'sensible' milkshake, with only 140 kilocalories. The researchers took blood samples from the participants at various time points to measure their ghrelin levels. In between the two blood draws, the participants had to examine and evaluate the (incorrect) milkshake label. They were then asked to drink the milkshake and then re-evaluate it. Analysis showed that the mindset of the people who believed they had been given a high-calorie milkshake resulted in a sharp drop in their ghrelin levels. The ghrelin levels of those who believed they had drunk a much lighter version of the shake remained virtually unchanged. Likewise, the physical reaction with respect to the hormone ghrelin was more strongly associated with the number of calories people thought they had consumed than with the actual calories. The researchers concluded from this that the effect of our food on our ghrelin levels is mediated by the mind. So our mindset can influence our response to food!

More than ten years ago, the Dutch professor Aart Jan van der Lelij and the Italian professor Ezio Ghigo made an interesting discovery. Along with their teams, they found that unacylated ghrelin, a sister hormone of ghrelin, could curb the hunger hormone ghrelin, and that together these two hormones have beneficial effects on metabolism, including the part that regulates how our body responds to the sugars we eat. The results of their research could provide the impetus for new drugs for people whose ghrelin levels are chronically elevated or for those who would benefit from improved glucose and insulin metabolism, such as people with diabetes.

GUT HORMONES ALSO COMMUNICATE WITH OUR BRAIN

There are even more appetite hormones. Very important hormones are produced not only in the stomach, but also further along the digestive tract at the start of the small intestine. Early in the twentieth century, there were already indications that there must be substances produced by the gut that could stimulate insulin production at a distance, in the pancreas. The biological benefit of this would be that, with the help of insulin, the absorbed sugars from the gut could then quickly be absorbed by the body cells. This research disappeared into a dusty desk drawer for decades, and only found its way back onto the research agenda in 1964 – that's the way things sometimes go in science. That year, scientific research confirmed (once again) that there are gut hormones present in the body. It turned out that taking glucose orally led to more insulin being released than when glucose was injected directly into the bloodstream. Further research showed that this must be mediated through hormones released by the gut. So what were these mysterious substances? The first hormone, gastric inhibitory polypeptide or GIP, was only discovered in 1970. This was followed in 1984 by glucagon-like peptide 1, or GLP-1, the second and also most important of these mysterious gut hormones. They are produced after a meal is eaten, and go by the collective name of 'incretins'.

In the meantime, a chemically modified form of GLP-1 is being used to treat diabetes (because it increases the amount of insulin released by the pancreas); in some countries it's also being prescribed in a slightly higher dosage as an anti-obesity medication. In contrast to ghrelin, the incretins curb appetite, which over time can lead to weight loss. That incretins curb appetite seems logical, seeing that, after a meal, signals have to be sent to the brain to tell you to stop eating. But how does the gut know it has to release incretins? What triggers this is a rise in the blood glucose level, which is exactly what happens following a meal.

After eating a meal, the gut hormones peptide YY (PYY) and cholecystokinin (CCK) are also released. These two hormones act in a way similar to incretins. One of the things that PYY does is slow down the emptying of the stomach so that you feel full sooner. This hormone also increases the amount of insulin released by the pancreas and curbs appetite. CCK slows the emptying of the stomach as well, and is also a satiety hormone (see **Figure 5**).

'CANNABIS' INSIDE YOUR BRAIN: THE ENDOCANNABINOID SYSTEM

Anyone who has ever tried marijuana will recognize this: in addition to the high, your appetite often increases, often known as the 'munchies'. Around five thousand years ago, people discovered the effects of cannabis, or marijuana. It was one of the first plants to be used as medicine, and also for recreational purposes and religious ceremonies. Thousands of years later, the active ingredient in cannabis was shown to be delta-9-tetrahydrocannabinol, or THC. The receptors that cannabis binds to were also discovered. And it turns out that our body has its very own endocannabinoid system.

Just like some other species of animals, humans make their own endocannabinoids. These are fat-like substances in the body that activate the same receptor as THC, the active ingredient in cannabis. These endocannabinoids are made only when they are needed, and play an important role in how we feel, our memory, and our brain's reward system, as in alcohol and drug use. Long-distance runners are also familiar with the effects endocannabinoids have on the brain: minor pains are suppressed, and euphoria can arise during physical exertion (the 'runner's high'). You can also experience this euphoria with a substantial amount of cannabis.

But why does cannabis increase appetite? Just like our body's own endocannabinoids, the THC in cannabis is able to bind with endocannabinoid receptors in the hypothalamus in our brain. The effect it has on appetite is mediated by receptors in the hypothalamus. If these are switched 'on', as it were, your appetite will increase. THC and the body's own endocannabinoids can also have an effect on our fat and sugar metabolism and energy balance. These effects are mediated by receptors that can be found on all sorts of body cells, including muscle and fat cells. The stimulation of these receptors on these organs leads to slower sugar and fat metabolism, and thus also to an energy balance that leans towards storing more fat. Moreover, it seems that the neurons in the hypothalamus itself also produce endocannabinoids that help to strictly regulate our feelings of hunger. In the body, little is left to chance, just like at a well-organized airport with a professional air traffic control system, which we compared our human appetite regulation system to earlier. However, sometimes one disruption will lead to another. If a plane arrives late, some passengers will miss their connecting flight. When body weight is too high, this almost always results in a number of disruptions. Research has shown that the hypothalamus of obese mice with no leptin produced abnormally high quantities of endocannabinoids, which contributed to increased feelings of hunger.

Since endocannabinoids can both stimulate our appetite and slow down our sugar and fat metabolism, the idea to develop a drug that would block this response was a logical one. And so it came to pass: at the beginning of this century, the anti-obesity drug rimonabant was developed. This drug blocks the effects of endocannabinoids on one of

81

its receptor types (because there are different kinds of endocannabin-oid receptors) and, when used in combination with a healthy lifestyle, was found to be effective for losing weight. The drug was enthusias-tically received because it suppressed the appetite, and this suddenly made it much easier for those taking it to reduce their food intake. Rimonabant was approved in Europe in 2006 and brought onto the market in thirty-eight countries worldwide, including Mexico and Brazil, but not in the USA. Unfortunately, in the years that followed, users of the drug not only lost weight, some of them also experienced severe psychiatric side effects. Among other things, people taking rimonabant were at greater risk of depression, and some of them even committed suicide! The drug was globally taken off the market in 2009. However, there is still a great deal of interest in endocannabinoids and their potential to combat overweight.

WHY DO WE ALSO EAT WHEN WE'RE NOT HUNGRY? THE ROLE OF THE REWARD SYSTEM

Think about a lavish Christmas dinner when, after eight copious courses, you still manage to eat a scoop of ice cream, a tartlet and a piece of cheese, even though all of the alarm bells in your hypothal-amus are going off to let you know you're full. So why do we keep eating? To be sociable? Or because your host has been slaving away in the kitchen for the past two days and you don't want to disappoint him? Because you learned you're supposed to clean your plate (in this case, plates)? Or – and this is usually the reason! – because the food simply tastes too good? This last phenomenon is something we also call 'hedonic eating'. This kind of eating is driven not so much by our physical need for food, but by the reward system in our brain. This system is genetically determined in part, but has also been shown to be learned to some extent.

Imagine that a child – let's call him Lucas – falls off his bicycle and hurts his knee. His mother hurries over and lovingly helps him up. She takes him inside, looks at the scrape and dries Lucas's tears. After deciding that a plaster would be pretty useless, she gets her son a nice sweet. When Lucas sees it, his tear-stained face brightens. The sweet comforts him. And, at that moment, Lucas's reward system is operating at full tilt, and he's taken another step towards learned behaviour: food can comfort you!

Behaviour researchers have identified three components of reward that come into play for future 'hedonist' Lucas: 'liking', 'wanting' and 'learning'. Although these components of reward are connected, they are driven by different mechanisms. The sensation of taste generated by contact with food ('liking') leads to the motivation to want to eat it ('wanting'). These components pertain to a reward's 'hedonic' impact (that is, the sensation of pleasure associated with a reward), and the motivation for a reward, respectively. 'Learning' pertains to the associations we attach to a reward. This is also how learned behaviour is established, as when Lucas got a sweet after scraping his knee. The sensation of pleasure produced by the sweet is then linked to the idea that having a sweet is a good way to alleviate pain or find comfort.

Tests in animals have shown that the endocannabinoid system is important to hedonic (sensory) experiences of food. The neurotransmitter dopamine – which is also referred to as one of the brain's 'happiness hormones' – plays a role in the components 'wanting' and 'learning'. Dopamine forms an important link in our reward system, and therefore also in addiction. When someone who is addicted drinks alcohol or uses drugs, he or she produces an enormous amount of dopamine. In non-addicts as well, dopamine produces wonderful feelings of pleasure and joy. For example, dopamine is released when you see or smell a freshly baked cake after a long walk through the wind and rain. Exposure to pleasant food-related stimuli – such as the taste, smell or appearance of food – activates our brain's reward system. Think of those stalls selling syrup waffles that you pass when you're out shopping in your local high street. Just the smell of those warm, sweet waffles is enough to provide you with a pleasant stimulus. The brain also responds in this way to stimuli such as music, money, sex and drugs. Research using special brain scans has shown that when people with obesity see photos of food, their brain activity is different from that of people with a normal weight. Because of these brain responses, it's not surprising that obesity is sometimes considered to be an addiction to food.

Another substance that plays an important role in addiction and can also curb appetite is serotonin. Serotonin is one of our 'happiness hormones'. It makes us feel good, is calming, helps us fall asleep, has an effect on our sensitivity to pain, and makes us emotionally stable. Serotonin also sends a signal to your brain (including via the MC4 receptor) that you feel full, even though your stomach may not be full. Eating carbohydrates also results in an increase in serotonin, and so some people like to eat lots of carbohydrates because this makes them feel good. Unfortunately, an overabundance of carbohydrates also contributes to obesity, so those good feelings are short-lived.

Lorcaserin – a drug that curbs appetite – has come onto the market in a number of countries; this drug is able to activate serotonin receptors in the brain. While it's clear that the parts of the brain involved in addiction also play a major role in eating, these relationships have not yet been sufficiently demonstrated to be able to call obesity a true 'eating addiction'.

Still, there are those for whom eating is an actual obsession because they feel extremely hungry all the time, as with Karen and Jack. And, just like with other addictions, in such cases the kitchen cupboards that contain food have to be locked as a preventive measure, and there's a chance that even the rubbish bin will be searched in a desperate attempt to find food. It is possible that, in the foreseeable future, a drug that can restore a feeling of satiety will become available to some of those who have this kind of rare genetic abnormality. The results of the initial studies into this are promising. But unfortunately, for some of these people, this endless feeling of hunger is not yet treatable. They will have to learn to cope with this extremely problematic disruption of their hunger hormone system, and learn to accept that their body will always be too heavy. And hope that others will accept them too, which is certainly not always easy in a society where people who struggle with severe overweight face discrimination.

Having said that, it's both important and useful for everyone to know how to feel full more easily. You can also use this knowledge about hunger and satiety hormones to keep the amount of food you eat every day within limits (see **Box 7** for handy tips for feeling full faster).

How can we promote healthy eating behaviours?

In many countries, you only have to step out your front door to find shops and stalls selling the most delicious foods. Just try to resist . . . And, if adults find this hard, what must it be like for children? In the world we live in, there's a lot to be gained from reining in our food intake.

Some countries have implemented drastic measures, like the 'sugar tax' (which makes products high in sugar more expensive), or through legislation that restricts marketing unhealthy products like sweets, soft drinks, cakes and snacks to children. Some measures have proven to be very effective. The WHO is advocating for a tax on beverages that contain sugar and for better food policies in schools. Forty countries have already introduced a tax on products that are high in sugar. In Mexico, for example, this measure has been successful. Mexicans are crazy about soft drinks. After the sugar tax was introduced, consumption of sugary drinks in Mexico dropped by nearly 8 per cent. Not only does the sugar tax contribute to a healthier lifestyle, in an ideal world, governments can then use the income this generates to take even more preventive measures and to create a healthy living environment.

But there are more opportunities for encouraging (healthy) behaviour than just legislation and tax measures. The Dutch behavioural scientist Roel Hermans has many good ideas about this. Portion sizes of high-calorie products could be reduced (not too small, of course – no one wants to buy a package with just three M&Ms), and portion sizes of healthy products could be increased. These adjustments respond nicely to the so-called portion size effect: the tendency for people to adjust their eating behaviour to the portion they are offered. In this way, people unconsciously eat more of the healthy products and less of the unhealthy products. Food manufacturers could adjust the size of their packaging and create a 'medium' that is more in proportion to facilitate this 'default effect'.

This effect can be put to use in creative ways. Take a milkshake at McDonald's: if the medium is made the size of a small, and the small is then made even smaller, people will still often choose the medium size. It also helps when larger numbers of the smallest size are stocked on the shelves, and a lot fewer of the larger sizes. Then people get the feeling that the smallest size is 'normal', and they choose this more often. Apparently, many people like to conform to the norm. It would be helpful if supermarkets made the prices of the smaller portions more attractive. It's usually the other way around: the larger the package, the cheaper the product costs per gram. That's fine for washing powder, but when it comes to unhealthy food, all too often this prompts people to make the wrong choice.

Box 7. Handy tips for feeling full faster

Some foods more than others send your brain a stronger signal to let it know you're full. Eating these foods or following these recommendations will allow you to feel full faster or for longer:

- Choose unprocessed food products as often as possible. These induce a normal decrease of the hunger hormone ghrelin and an earlier satiety signal – in contrast to ultra-processed food from the supermarket.
- Add an egg (contains lots of protein), oatmeal (rich in fibre and protein), or some unsalted nuts (in addition to protein, they also contain healthy unsaturated fatty acids) to your breakfast so that you feel full until lunch.
- Try adding half an avocado or some legumes (chickpeas, lentils or beans) to your lunch.
- Have a cup of low-calorie soup or a glass of water (preferably cold!) before meals.
- Try using fermented products (like pickled gherkins and sauerkraut) and chilli powder ($\frac{1}{4}$ teaspoon with your meal), which also stimulates your metabolism.
- Since it takes an average of twenty minutes for your satiety hormones to kick in, it helps to eat slowly and chew your food for a long time. Pay attention to how often you usually chew (four times? eight times?) before swallowing your food, and try doubling or tripling this.
- Eat from a smaller plate (the optical illusion helps you feel fuller) and with smaller utensils (this helps you eat more slowly because there is less food per bite). In this way, you'll give your body more time to produce the satiety signals.

- Ideally, portion out the food onto the plates in the kitchen. Or, if you're serving the food at the table, right from the pan, make sure it's just enough so that the aroma of food in the pan doesn't keep whetting your appetite.
- Eat mindfully! In other words, not in front of the TV. Instead, use your senses to alert your brain to the fact that food is coming in, so your satiety system can be activated on time.

Research has shown that people eat 35 per cent more if their portion size is doubled. Such a large portion seems to provide an unconscious visual stimulus to eat more. And this applies to an even greater extent to unhealthy food than to healthy food. Think of an all-you-can-eat buffet where an abundance of food just sits there beckoning you. There's a good chance that on such an occasion you'll eat more than when you portion out food on your plate at home. In addition, you're more likely to take bigger bites from a larger serving. This means that you have less 'mouth contact' with the food you're eating; this mouth contact is what you need to generate a feeling of satiety in your body. So the food in one of those three-Michelin-star restaurants, where you get a huge plate with a tiny mound of beautifully presented food that you eat in little bites at a leisurely pace, is actually ideal for generating a solid sense of satiety and consuming fewer calories. But alas, these kinds of meals are pricey, and your bank account will be empty before you feel full.

In short, although we have appetite and satiety hormones that can make us feel hungry or full, we can overrule these signals with processes that are mainly driven by the mind – for example, because we simply find something very delicious and experience a sense of reward when we eat it. As a result, we can still manage to put away that dessert of warm apple cake with whipped cream in spite of the fact that we feel full after eating a three- or four-course meal. The tips in **Box 7** will hopefully make it easier for you to hold your own in this world filled with tempting food.

6

Marvellous metabolism

OUR AMAZING INTERNAL COMBUSTION ENGINE

You're at a birthday party and you told yourself you were only going to have one small piece of cake. Just a taste. Summer is on the way . . . The person sitting next to you is enviably thin, and you think, 'That poor soul must live on cucumbers'. However, nothing could be further from the truth. First, your neighbour eats a piece of cake that's twice as large as yours, then has two glasses of cola (not the diet version), and finally they polish off a whole pile of crackers and spreads! A person like that must have to run a marathon a week to stay so slim! You discreetly sound out your neighbour and ask whether they play a sport. Their reply astounds you: 'I don't like playing sports at all, except for going on a short walk with the dog when I take him out. But sports? No . . .'

How does a person who doesn't exercise and eats like there's no tomorrow stay so slender? Welcome to the wonderful world of our metabolism! Many people think of this system as a kind of internal combustion engine, like the one in a car, that can be turned on or off, but which burns fats and sugars as fuel. But the system is by no means that simple. People are more complex than cars . . . and combustion in a car engine is already quite complex. So, we're going to give you a crash course on metabolism.

Whether you're running a marathon or lounging in front of the TV, every cell in your body is constantly burning nutrients (particularly fatty acids, as we saw earlier) and converting them to energy-rich substances and heat. Gut cells, nerve cells, fat cells, muscle cells – every cell contributes to our total metabolism, although some cells contribute more than others. For example, if you're running a marathon, the cells in your muscles and lungs are working harder, whereas if you've just eaten a meal, your gut cells are the ones working overtime. Because every body cell contributes to metabolism, the total daily energy expenditure is divided into 'components'.

The first component is 'resting metabolic rate', or resting metabolism. This is all of the energy you need to keep your body going when you're at rest. Because your heart has to keep beating, your body has to be kept warm, your brain has to stay switched on and your hormone levels have to be maintained, this resting state alone is responsible for nearly 60 per cent of the energy you expend every day. Resting metabolic rate in women is around 1,400 kilocalories per day, and in men this is around 1,800 kilocalories. This is simply because, on average, men are larger than women and have more muscle mass.

To keep your metabolism going, first of all you need food, and of course it takes energy to digest food, absorb the nutrients and store any surplus fats and sugars. We call this mechanism the 'thermic effect of food', and this easily accounts for 10 to 15 per cent of your total energy expenditure. Indeed, you burn calories just by eating, although it will take more to burn a stalk of celery than a bowl of pudding because of the complex structure of the celery.

And, finally, there is 'activity thermogenesis' (thermogenesis = energy expenditure), that is, the amount of energy required for the activities you undertake, like walking, talking, working and exercising. This amounts to around 25 to 30 per cent of your total energy expenditure.

IN MOTION: STANDING AND FIDGETING

This activity thermogenesis may well hold the answer to the question of why some people seem to be able to eat endless amounts of food without gaining a bit of weight, even though they hardly seem to exercise. The emphasis here is on 'seem', because many people are more physically active than they think. Compare colleagues Chris and George. Chris is in his mid-forties and is slightly overweight. He has an office job, and sits at his desk all day long. The only exception to this is when he walks to the canteen for a sandwich during his lunch break. If he has a question for a colleague, he would rather send an email than walk over to their office. His colleague George is in his early fifties and looks fit. George installed an app on his phone that reminds him every forty-five minutes to stretch his legs and walk around for ten minutes. He gets a glass of water and then walks over to discuss something with a colleague. He's also got a special desk so he can work standing up.

Although Chris and George do the same kind of work, George burns a lot more calories every day than Chris does, and this happens almost imperceptibly. Sitting at his desk, Chris burns around 80 kilocalories an hour, while George easily burns 100 kilocalories an hour as he stands at his. George's body is also larger and has more muscle mass, which means that his resting metabolic rate is higher. When you add to this all of the additional calories burned during other physical and mental activities, by the end of the day, the difference between Chris and George in terms of energy expenditure will have quietly increased by a considerable degree.

In our Western society, people spend on average more than twelve hours a day sitting down, mainly in front of the computer or TV. When you add to this the seven hours we on average spend sleeping every night, people spend as many as nineteen hours a day either sitting or lying down. That's a whole lot more than our ancestors ever did . . .

And that's not good news, because people who sit much of the time are generally heavier, are more likely to have diabetes and are at greater risk of cardiovascular disease. So how can you compensate for this sedentary way of life? By exercising every day? Or are there other ways?

Researchers in the Netherlands studied this a few years ago. Three groups of research participants took part: one group (the control group) had to sit for fourteen hours a day. Another group had to sit for thirteen hours and replace the final hour with an hour of exercise. The third group had to replace six of the hours spent sitting with four hours of walking and two hours of standing. To see which group had the most favourable metabolism, the research participants were asked to drink a sugary beverage, and then measurements were taken to determine which group was best able to process the glucose in the drink. It will come as no surprise that the 'full-time sitters' didn't win, but what was surprising was that the group that had replaced sitting with walking performed better than the group that had replaced sitting with exercise. A difference was noticeable after only four days. So this study has shown that one hour of exercise every day cannot undo the adverse effects of an entire day of sitting. Other studies have also confirmed that standing for a few hours instead of sitting, and regular short walks, make you healthier. Moreover, moving around in between other activities has been shown to improve your mood. So, try raising your desk so you can work standing up from now on (this will also help prevent you developing back problems). If your colleagues give you strange looks, tell them this: 'Winston Churchill always wrote his speeches standing up.'

Take it from us: you don't have to run marathons to be making healthy progress (see **Box 8**). In addition to working standing up and taking regular short walks, there are countless other, more subtle, ways to burn extra calories almost without noticing. For example, rocking your feet back and forth, or continuously playing with your keys, paper clips or other office supplies. We call people who do these kinds of things, often without realizing it, fidgeters. Research into this phenomenon has shown that people who are slender often turn out to be more fanatical fidgeters than those who are overweight. By doing nothing more than fidgeting, they can sometimes entirely reverse the harmful effects of too much sitting. So, if you're looking for a simple way to lose weight, buy a stress ball (it doesn't make any noise) or tighten the muscles in your buttocks regularly (no one will see).

Box 8. Tips for boosting your metabolism through exercise

By being more physically active every day, you can boost your metabolism and burn hundreds more calories every day. Some tips:

- Do you have a sedentary profession? Work a few hours every day at a standing desk and try to take a short walk once an hour.
- Use a pedometer and try to take at least 10,000 steps every day.
- Try becoming a fidgeter: consciously tighten your muscles now and then, tap with your pen, or fiddle with something.
- Get a dog!

A WARMER-UPPER: BROWN FAT

Standing up, walking around, fidgeting – these are all good ways of boosting your daily energy expenditure. But there's yet another way to convert calories into heat, because that's actually what metabolism is.

This has everything to do with our brown fat. That's right: we have two kinds of fat in our body, white and brown. The body fat we've been talking about up till now, and which stores fat, is 'white fat'. Both kinds of fat owe their names to – you guessed it – the colour of the fat. The first evidence for the existence of brown fat dates from around 1551. At the time, Swiss naturalist and researcher Konrad Gessner described brown fat as 'neither fat nor flesh [nec pinguitudo, nec caro], but something in between'. And Gessner had a good point: brown fat is an organ, and in many respects seems to be a cross between white fat and muscle.

Biologists have long known that large quantities of brown fat are present in animals that hibernate during the winter, such as hedgehogs. When they hibernate, hedgehogs live off their fat reserves the entire time. By keeping their body temperature as low as possible, they use a minimal amount of energy during this period. Hedgehogs' body temperature can drop to near freezing. Just before they come out of hibernation, their body has to warm up fast. To do this, they make use of a little heater inside their body that can rapidly convert fats and sugars into heat, and the hedgehog's body warms back up. This heater is brown fat.

Human babies also have a large amount of brown fat, especially between their shoulder blades. This fat is essential to babies because they lose a great deal of heat through their relatively large head, and they don't yet have enough muscle to produce heat by shivering (because, when you shiver, your muscles produce heat). By stoking their little brown fat-store heaters, babies stay nice and warm.

Brown fat is a fantastic remnant of evolution. Our prehistoric ancestors lived off of their white fat during periods of famine. But they also lived through periods of extreme cold during ice ages. Because shivering uses up a great deal of energy, when times are hard it's very useful to have an extra organ that can warm you up from the inside out. It's been a very long time since the last ice age, and nowadays our houses are pleasantly warm. So, from puberty onwards – when our muscle mass increases and we're able to shiver effectively – we're less dependent on our brown fat, and the brown fat between our shoulder blades largely disappears once we leave childhood. For a long time it was thought that this fat disappeared for good, but recent research has shown that this is not entirely true . . .

The (re)discovery of brown fat was in a certain sense a lucky coincidence. Nuclear medicine departments in larger hospitals use what are known as PET (positron emission tomography) scans to look for cancer. To be able to detect cancer cells, a radioactive substance that resembles sugar is injected into patients' blood vessels. Because cancer cells have a higher metabolic rate and absorb lots of sugar, they look brighter on a PET scan than healthy cells. However, around fifteen years ago, the nuclear medicine physicians noticed something odd while looking at one of these scans. In a number of patients who had been scanned during the winter months, they observed an uptake of sugar in strange places, such as in the neck and around the aorta. These were very unlikely places for cancer. What was going on? They decided to take a biopsy of the tissue that had shown up on the PET scan. When this was placed under the microscope, the researchers couldn't believe their eyes: they saw tissue filled with tiny droplets of fat and packed with little 'power stations' known as mitochondria. This also contained a special protein found only in brown fat. To their surprise, they had (re)discovered brown fat! Many studies have since shown that adults still have brown fat.

When you expose adults to warmth and then do a PET scan, the scan shows that the brown fat absorbs little or no sugar. This is logical, because brown fat doesn't need to be active if it's warm. But when you expose these same people to mild cold for two hours (15 to 17°C is cold enough), entire bands of active brown fat can be seen deep inside the body, especially along the aorta and in the neck (see **Figure 6**). The younger and leaner a person is, the more brown fat there is. Young adults are estimated to have around 300 grams of brown fat. Of course, this is next to nothing compared with the amount of white fat (which can amount to tens of kilos). But with the (re)discovery of our brown fat, an exciting new research area was born.

BROWN FAT BURNS FATS TO PRODUCE HEAT

Does the discovery of brown fat provide the solution for people with over-weight? To answer this question we must first have a closer look at exactly how brown fat works. For example, how does brown fat 'know' that it needs to switch itself 'on' and generate heat? It gets this information by way of a very ingenious system. Everywhere in our skin are temperature

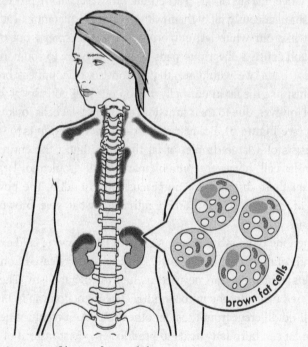

Figure 6. Location of brown fat in adults.

sensors that register whether the skin is being exposed to either cold or heat (just like the thermostat in our living room). These sensors in the skin pass this information on to the temperature centre in our brain. This temperature centre is located in the hypothalamus, the control tower in the brain we mentioned earlier, which also regulates metabolism and appetite. It processes all of the incoming information and decides whether we need to stoke the fire or get rid of some of the heat. The latter is achieved by dilating the blood vessels in the skin so that you start to sweat. You notice this, for example, when you go to the sauna: because your blood vessels dilate, your skin gets flushed and warm. When the heat has to be turned up, the brain sends a signal to the brown fat that switches it 'on'. This happens via certain nerves and takes just a few seconds. Once this signal has reached the brown fat, many processes take place simultaneously, all of them with the aim of generating heat.

Just like our white fat, our brown fat is also spread out over our body and consists of various 'pads' of fat. If we were to zoom in on one of these pads, we would see that it consists of countless brown-fat cells, just like the fat in our belly and on our hips consists of white-fat cells. However, due to their function, the brown-fat cells look very different (see **Figure 6**). Because a white-fat cell's function is to store fat, it consists of a large droplet of fat that fills almost the entire cell. A brown-fat cell, however, contains many small droplets of fat, and in between these droplets are numerous mitochondria, the power stations of the cell. These mitochondria are what give brown fat its brownish colour.

When the brain switches on the brown fat following cold exposure, all kinds of things happen at once. Fatty acids are released from the fat droplets, and these can be used as fuel for metabolism. These fatty acids are burned by the mitochondria. The mitochondria in brown fat are a little different from those in other cells. As a result, if necessary, brown fat can burn fatty acids to produce only heat instead of energy-rich substances.

CAN BROWN FAT HELP YOU LOSE WEIGHT?
BARBARA'S STORY

But does this also mean we can lose weight by making our brown fat work harder? Yes, we can, as Barbara's story will show.

Barbara is a sixty-one-year-old woman who works in a clothing store. She is married and has two adult daughters. In her free time, she does yoga and likes to cook. Barbara had been slender her whole life, and kept a close eye on her weight to make sure it remained stable. And then something strange happened. Barbara: 'I noticed I was getting more and more hungry, and that a sandwich at lunch wasn't enough. During the afternoon, I'd get so hungry I'd raid the biscuit tin in the pantry. My colleagues even started to notice. They knew I always watched my weight. I would also take an extra helping at supper, and, even then, I would be hungry again by bedtime.'

But, in spite of her increased appetite, Barbara didn't gain weight. In fact, she started losing weight. After three months, the scales showed she'd lost 5 kilos. 'When I would look at myself in the mirrors in the clothing store, I could see that my clothes were starting to hang on me, and that my face was getting gaunt. Regular customers started to ask about it. "Are you sick? You look so thin." And sometimes I also felt strange. I would get so hot for no apparent reason!' Barbara decided to give it some time, but when after a month she'd lost yet another kilo, she decided to see her general practitioner, who also didn't know what to make of Barbara's story. The blood tests that followed showed nothing out of the ordinary. 'My thyroid was working properly, and my other blood tests also came back normal.'

Barbara was referred to the hospital, where she was examined to see if there were signs of an inflammation or if there was another reason her metabolism was so fast. Could it be a tumour? 'I started to get really worried. What was wrong with me? A few days later, I had a PET scan, which showed something unusual.' Near her hip was a round mass, 6 centimetres in diameter, that was absorbing a large amount of sugar. Was it an inflammation? Or a tumour, after all? They took a biopsy to determine what kind of cells the mass was made up of. The results came back quickly: the cells were filled with fat and mitochondria. It was a benign tumour made up of brown-fat cells, something that is very rare. Once the mass was removed, Barbara felt much less hungry and she gained 10 kilos in the months that followed!

From Barbara's story we can see that, when your body contains a substantial amount of extra brown fat, you rapidly shed lots of kilos. Although her story is exceptional, this 'side effect' of brown fat is precisely why scientists are so excited about it. Is it possible to boost metabolism by stimulating the brown fat already present in the body? Even if this is only 300 grams?

Researchers from the research team Mariëtte Boon is affiliated with, at Leiden University Medical Center, wanted to gain insight into the extent to which brown fat contributes to metabolism in healthy individuals. To do this, they conducted an experiment in which they exposed healthy young men to acute cold by having them lie between special mattresses with cold water flowing through them. They measured their metabolic rate before and at the end of the cold exposure. Cold, after all, is the natural stimulus for activating brown fat. After just two hours, these men's metabolic rate increased by an average of 200 kilocalories a day. So, what the study showed is this: if you 'switch on' that remaining bit of brown fat and turn it up as high as you can, you can burn an extra 200 kilocalories a day, which would amount to burning 8 kilos of fat a year. This may not seem like much, but, for people who are severely overweight, losing 8 kilos of fat would already have many positive health benefits. For example, their body would become more sensitive to the effects of the hormone insulin (which would result in a drop in their blood sugar levels and so their risk of diabetes would also decrease), and a decrease in the level of fats in the blood would result in less accumulation of fat in the liver.

A group of Japanese pioneers in the field of brown fat research decided to put this to the test and they thoroughly investigated brown fat's potential to improve metabolic health. Ten healthy young men were prepared to take part in a study that would happen in the cold. During this study, the men were exposed to a temperature of 17°C (so, just below normal room temperature), two hours a day, for six weeks; this was not required of the control group. And what did they see? The group that had been exposed to cold lost nearly a kilo of body fat over the course of those six weeks.

Of course, healthy young men aren't the target group of this research – you really want this to work in people living with obesity and diabetes! A study by a group of researchers at Maastricht University in the Netherlands has already shown that, in men living with obesity, the amount of brown fat in their bodies increases if they undergo a ten-day 'cold treatment' (six hours a day, in shorts and a T-shirt, in a room with a temperature of between 14 and 15°C). This short cold treatment also had extremely positive effects in men with diabetes. As a result of this same spartan cold protocol, these men became much more sensitive to insulin and were able to nearly halve the amount of insulin they needed to inject. This could probably be attributed in part to the beneficial effects of brown fat. Unfortunately, how long this positive effect lasted was not studied. What we have learned from these studies is that brown fat has the potential to speed up our metabolism and reduce our (white!) fat mass. There are a few simple tricks you can use to train your brown fat (see **Box 9**).

DON'T LEAVE ME STANDING IN THE COLD!

Most people don't really like to be cold, and in the winter they'd rather curl up under a blanket on the sofa (something that does not activate your brown fat). Consequently, a great deal of research has been done into ways to activate brown fat without having to stand in the cold – for example, by using hormones, food or medication. For now, this research is being carried out primarily in mice.

Box 9. Tips for boosting your metabolism using your brown fat

Brown fat converts calories into heat, and in this way can help you lose fat. Because brown fat is so sensitive to cold, there are simple ways to 'train' your brown fat every day. You can do this in the following ways:

- use cold water for the last few minutes of your daily shower
- take a cold bath occasionally
- turn down the heat by a few degrees for a few hours every day, without putting on a sweater
- have children play outside without a jacket more often (children don't get colds from the cold but from a virus!)
- exercise outside rather than in the gym and cycle to work, especially when it's chilly
- eat red pepper and drink coffee and green tea

There's nothing wrong with being cold once in a while!

An example of a hormone that can activate brown fat is the thyroid hormone, which we mentioned earlier. This hormone can penetrate brown-fat cells and increase metabolism in various ways through the brown fat. This is probably one of the reasons that people with an overactive thyroid often feel hot and also lose weight, even if they're eating more. The opposite is also true. For brown fat to function properly, thyroid hormone is essential. This is why people with an underactive thyroid feel cold, even though the ambient temperature is within a normal range, and also why they gain weight.

Examples of compounds in foods that activate brown fat in mice are capsaicins (a compound found in hot peppers), catechin (found in green tea) and caffeine. The list of medications that can potentially activate brown fat is getting longer by the week. Antidiabetic drugs, ADHD medication and even certain kinds of bladder medication have been shown to increase metabolism in mice through brown fat. And there is yet another way to increase brown fat activity. Studies in mice have shown that, under certain conditions (for example, following cold exposure or due to certain medications), white-fat cells can turn into brown-fat cells – almost like magic! Just imagine. We've got such a huge number of white-fat cells, if we could change just a fraction of these to brown-fat cells it would give our metabolism a considerable boost. Some researchers have even taken this a step further: liposuction could be used to remove a little white fat, which could be cultured in a Petri dish. The white-fat cells could then be treated with a substance that would transform them into brown-fat cells. And this brown fat could then be reintroduced back into the body to increase the total amount of brown fat. Although this might sound like science fiction, it has already been done successfully in mice! So, who knows . . .

Unfortunately, the results from mouse studies are not always directly applicable to humans. After all, a mouse is not a human. Also, because of their compact bodies, mice have much more brown fat than humans in relative terms, which is why a therapy that activates brown fat will have a much greater effect in mice. For example, when the brown fat of mice is activated, they can lose half of their total body fat within a few weeks. In addition, their blood sugar level and blood fats like cholesterol drop drastically, and it can even reduce atherosclerosis.

Even so, some interesting results have already been obtained in humans. For example, capsaicin – the compound found in hot peppers – speeded up metabolism in healthy young men when they took this in pill form for six weeks, possibly because their brown fat became more active. And many drugs are currently being tested on humans, with promising initial results. So, there are all kinds of developments taking place around brown fat. That cold is effective at stimulating brown fat is beyond dispute; but the long-term effects of various drugs and foods still need to be demonstrated in the coming years. So, until then: eat spicy foods and cool things down with a cold shower!

7

Fat and our biorhythm

A SERIOUSLY DISRUPTED BIORHYTHM: FRANCESCA'S STORY

Francesca is a flight attendant. Although her job is enjoyable and varied, it's also sometimes turbulent – both literally and figuratively – and very bad for her natural day/night rhythm. 'I remember well how, once, during a flight to New York, we encountered a huge amount of turbulence, and the plane suddenly dropped a couple of metres.' The 'fasten seat belt' sign came on as seat backs jolted and laptops clattered. Cups rattled, and a grey-haired man spilled hot coffee on his suit. Francesca heard people cursing. A toddler started to wail, and her mother hurriedly searched for a dummy. Francesca was not easily phased. Gracefully tossing back her long blonde ponytail, she walked resolutely over to the flustered man with the spilled coffee and deftly cleaned his suit. She smiled patiently, and made sure that those in the rest of the cabin also came through the unexpected severe turbulence well. 'Luckily, the turbulence finally passed and we landed safely in New York.'

A couple of hours later, Francesca was in her hotel room at last. Although she was exhausted after the flight, she was also completely disoriented because, according to her internal clock, it was late at night, but the sun was still shining. Francesca: 'I kicked off my shoes, and collapsed onto the king-size bed. Phew. That flight was behind me. I wanted to go to sleep. I was soooo tired, and I had a headache. I felt a little sick. I couldn't tell if I was hungry, or not hungry at all. Luckily I found a bar of chocolate in my bag, and that went down a treat. I told myself I was just going to have one more piece, but, before I knew it, I'd wolfed down the whole bar in one go. A few days before the flight to New York, I'd returned from another intercontinental flight, to Bangkok. I wasn't really sure which time zone my body was living in at that moment. Here, in New York, it was only four in the afternoon. Now what? Stay up and go have some supper first? Or just surrender to the delicious hotel bed and crawl in between the snow-white sheets?'

Anyone who has travelled to another time zone or worked night shifts will recognize themselves in Francesca's story. The complete disorientation, wondering whether you should go to sleep or not, the overwhelming fatigue, and especially the craving for high-fat food that you are more than happy to yield to. When you disrupt your body's biorhythm, this does not go unpunished. In recent decades, the number of people working nights shifts worldwide has increased substantially. It is estimated that 15-20 per cent of workers in industrialized countries are employed in shift work. People who work night shifts are more often overweight and are at greater risk of developing a range of diseases, including type 2 diabetes and chronic kidney disease, and potentially some forms of cancer as well. Why is this? What happens to your body fat and your appetite when your biorhythm is disrupted? And how does your body know what time it is?

OUR BIOLOGICAL CLOCK

When we look at the universe, all sorts of things have rhythms – the turning of the earth and the other planets, the seasons, the days, the trees and plants, the animals – and the human body as well. Since the eighteenth century, it has been known that organisms have a kind of internal biological clock that governs the rhythm by which they live. At the time, French astronomer Jean-Jacques d'Ortous de Mairan was studying a species of plant, and he noticed that the leaves of this plant opened when it got light and closed again when it got dark. When he placed the plant in a dark cupboard, the leaves continued to open and close at set times. Apparently, the opening and closing of the leaves was not dependent on light and was regulated from within the plant itself by a kind of internal clock.

For a long time, we didn't know quite how the biological clock worked or whether people had one too. Important discoveries have been made in this area in recent decades, and these discoveries were ultimately awarded the Nobel Prize. In the 1970s, the American researcher Seymour Benzer and his student Ronald Konopka discovered that abnormalities in an unknown gene in fruit flies completely disrupted their biological clock. They called this unknown gene 'period', and the protein this gene encodes was called 'PER'. In the 1980s, other American researchers continued to work with this gene. They successfully isolated the gene, and discovered that PER accumulated in the cell nuclei at night and broke down again during the day. Amazingly enough, the PER protein levels turned out to fluctuate in twenty-four-hour cycles, completely in synch with the human sleep/wake cycle. But there were still some pieces missing from the puzzle. What curbed production of the PER protein in the cell nucleus at a given moment so it didn't produce PER continually?

Shortly thereafter, another American researcher, Michael Young, found the answer to this question. In addition to period, he also discovered two 'clock genes' that worked together with the period gene, and he gave them the fitting names of 'timeless' and 'doubletime'. These genes enable the PER protein to penetrate the cell nucleus and suppress the activities of the period gene so that the PER protein does not accumulate in the cell nucleus in excessive amounts. This fantastic system is what gives the proteins inside the cells their twenty-four-hour rhythm. The research team – made up of Michael Young along with Jeffrey Hall and Michael Rosbash – thus discovered the mechanisms of the biological clock and the way in which the human body synchronizes with nature. This was a major scientific achievement, and they were awarded the 2017 Nobel Prize in Physiology or Medicine.

More and more has become known about the biological clock of humans since the discovery that everything in nature has its own rhythm. This biological clock largely determines our sleeping patterns, hormonal activity, body temperature, blood pressure and eating behaviours. In this system, the hypothalamus was found to play the role of master clock. So, in addition to its 'control tower' function of releasing and receiving hormones and also regulating appetite, metabolism and body temperature, it turns out that the hypothalamus does so much more. Think of the hypothalamus as a big classic wall clock hanging in the living room that ticks according to a fixed rhythm and sets the tempo for the rest of your body. Not so long ago, various organs were discovered to have their own clocks. You can think of these as alarm clocks on bedside tables and small wall clocks in the other rooms of the house, which are all connected to each other in a wonderful way and synchronized by the large clock in the hypothalamus (see **Figure 7**).

HOW DOES OUR LIFESTYLE DISRUPT OUR BIOLOGICAL CLOCK?

Our eating pattern affects our biological clock. This becomes clear if we disturb our rhythm and, for example, eat large amounts of food in the middle of the night. Then all of the little clocks in the various organs get confused, and their connection to the 'master clock' is disrupted. Eating in the middle of the night happens more often than we might think. Of course, there are people who suddenly get hungry in the night and might occasionally have something to eat. But think

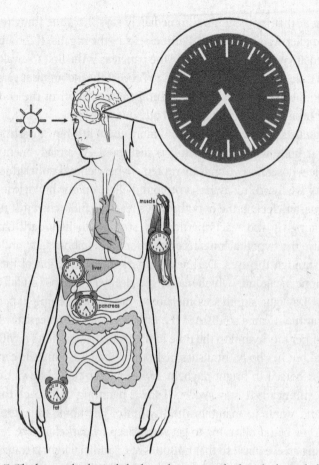

Figure 7. The human biological clock, with a master clock in the hypothalamus and smaller clocks in various organs.

about all of those people who work night shifts or, like flight attendant Francesca, travel through multiple time zones. They have to eat during the night. Eating during the night disrupts metabolism, and things like night-time blood sugar will also increase. That large master clock in the hypothalamus tells the body that it actually shouldn't be eating at that moment, and immediately says the same thing to all of the smaller wall clocks and alarm clocks in the organs. If food is consumed anyway, the little clock in the pancreas is the first to awaken, so it can quickly start making insulin to process the incoming sugars. And this immediately disrupts the natural rhythm of all of the countless other little clocks, bringing your body out of balance.

Scientists are learning more and more about just how disastrous it is for our body when this clock gets disrupted. As already mentioned, people whose biorhythm is disrupted are heavier and unhealthier. The society we live in today has a profound effect on our biological clock. The master clock in the hypothalamus reacts to more than just nocturnal eating. It also reacts particularly strongly to light/dark changes, because the hypothalamus receives information about light and darkness through the eyes. Darkness triggers the production of the sleep hormone melatonin, which makes you sleepy and helps you fall asleep easily. Daylight suppresses melatonin production. Being exposed to light at night can also disrupt the master clock. If Francesca wants to adjust her biorhythm to the time in the United States, if it's daylight on arrival but her body thinks it's night time, the best thing she can do is expose herself to bright daylight so her master clock knows that her body still needs to stay awake. If she is planning to fly back the next day and wants to maintain the rhythm of her home time zone, she would be better off trying to get some sleep by darkening the room or wearing an eye mask so that no sunlight or artificial light is transmitted as a signal to her brain through her pupils. 'Make sure the room is dark' is one of the first tips given to people who have trouble sleeping, along with making sure your bedroom is restful and well ventilated and with no TV or things that remind you of work nearby, no mental or physical exertion just before bedtime, no naps too close to bedtime, and moderate alcohol consumption.

One of the hormones that reacts to a disruption in the sleep/wake cycle and that has a strong biorhythm is the hormone cortisol. We also know cortisol as the 'stress hormone', because it is produced in greater quantities when there is stress (more about this later). We need cortisol day in, day out for a wide range of processes in our body, things like sugar metabolism and keeping our immune system functioning properly. At the start of every new day – between four and six a.m. – a generous shot of this hormone is released into your blood, and it peaks just before or while you are waking up, at around seven or eight a.m. You could say that cortisol is a kind of 'wake-up hormone'. During those first hours after waking (so, after the cortisol peak), the cortisol level drops, sharply at first, then more gradually over the course of the day, with another small peak during or after lunch. The cortisol level is at its lowest point between midnight and three a.m. This low cortisol level is very important for good-quality sleep. This system is precisely calibrated – for example, have a look at what happens when you 'skip' a night (disrupt your usual routine) and create what is known as 'social jet lag'.

LACK OF SLEEP MAKES YOU CRAVE SNACK FOODS: ERIK'S STORY

Erik had spent a very enjoyable evening with friends. He intentionally drank only one beer, because the following day was going to be a busy one at work, and he wanted to have a clear head. Things got completely out of hand anyway, even though he limited himself to that one beer. In the end, he spent the entire summer evening hanging out on the veranda with his friends, sitting in an old porch swing, talking football, telling questionable jokes, and laughing loudly more than anything else. It got late, certainly for a weeknight. Around midnight, their friend Johan, who lived next door, turned on his deep fryer to fry up some chicken nuggets. The platter of chicken nuggets was greeted with cheers and was empty in no time. At around half past two, Erik and two of his friends left on their bicycles. On their way home, they passed the snack bar, and Erik's mouth started to water. When he got a whiff of the delicious smell of deep-fried foods, his stomach cried out for yet another greasy snack. They got off their bikes and feasted on kebabs, and Erik also bought a chocolate bar. Yum. Although his stomach was completely full, he still managed to eat his 'dessert'. When he got home and looked in the mirror, he saw he had chocolate in the corner of his mouth and garlic sauce on his chin. Luckily, he succeeded in getting into bed without waking his girlfriend; he didn't dare give her a kiss with his garlic breath. Next morning, after just under three and a half hours of sleep, Erik was not exactly rested. He felt his heart pounding in his chest, and had a nagging pain inside his head, almost a crackling sensation. After a refreshing shower, he opened the fridge and poured himself a big glass of chocolate milk, then slathered chocolate spread over four thick slices of white bread. Erik was ready to start the day!

HOW TOO LITTLE SLEEP CAN MAKE YOU FAT

Lack of sleep gives you an appetite! And not for just anything. No, you're hungry specifically for high-calorie foods – in other words, you crave snack foods. So Erik's urge to eat a kebab after a night out is an entirely biological phenomenon. And research has shown that even a single night of fewer than five hours of sleep (measured as 'time in bed') disrupts our hunger hormones. In previous chapters we explained how the hunger hormone ghrelin stimulates your hunger, and that leptin provides you with feelings of fullness. Lack of sleep results in an increase in ghrelin and a decrease in leptin, and this is how not sleeping enough makes you hungry. Lack of sleep also makes your cortisol level rise, and this can make you hungry too. All of these things makes it easy to overeat, especially unhealthy foods. If Erik were to make a habit of these kinds of evenings that stretch into the wee hours, the lack of sleep along with eating at the wrong time (during the night) could even lead to considerable weight gain. You might think that, because you're awake and also active longer, you would automatically burn more calories. While this is true, the increase in the number of calories burned is generally not enough to cancel out the number of calories you consume as a result of your increased snack cravings.

Sleep deprivation is also known to lead to all kinds of other health problems. People are more emotional, concentration and memory decline, reaction time increases (and, as a result, lack of sleep leads to more accidents), susceptibility to infections increases, and the skin ages more quickly. But less well known are the above-mentioned effects sleep has on your appetite and metabolism. It would seem, then, that lack of sleep can also contribute to becoming overweight. What's more, it is known to cause blood sugar levels in the body to rise because the cells become less sensitive to insulin, and this can be the first step towards developing diabetes later on.

And, indeed, epidemiological research has shown a clear relationship between fewer hours of sleep and conditions like obesity, diabetes and cardiovascular disease. This correlation can already be seen when sleep duration is shorter than six or seven hours. This should not be underestimated, because a large poll survey conducted by the National Sleep Foundation in the United States has shown that one in five of those who participated in this sleep study slept less than six hours on weeknights. There are more numbers available about the relationship between hours of sleep and weight, such as those from a long-term sleep study that followed two groups of women for sixteen years. One group said they slept seven hours a night, and the other group less than five. Over the course of those sixteen years, the group of 'short sleepers' gained more weight on average than the 'long sleepers'.

Conversely, researchers from King's College in London found that more sleep resulted in better food choices. By the end of the study, it turned out that people who had been advised to sleep an hour and a half longer (and who actually did so) had reduced their intake of added sugars by 10 grams. They were also eating fewer carbohydrates per day than people who had not been given this sleep recommendation and had not started sleeping more.

Another large British study that followed more than 100,000 women showed that, ultimately, women who didn't sleep in bedrooms that were completely dark were heavier than women who did. This was not thought to be related to the appetite hormones, but to the master clock becoming disrupted by incoming light, which in turn disrupts metabolism.

Though it might seem far-fetched, to make it easier to stick to a healthy diet, you also have to make sure you get enough sleep. In recent years, a number of studies have shown that the relationship between too little sleep and overweight cannot be attributed only to a change in the appetite hormones leptin and ghrelin. There are many more factors involved. One of these factors is the quality of our sleep, and this quality might be even more important than how long we sleep.

Obesity is a major sleep disrupter. People with obesity often have 'obstructive sleep apnoea syndrome' (OSAS), which Rob in Chapter 4 also suffered from. This is characterized by loud snoring alternated with pauses in breathing (called apnoea), during which the oxygen level in the blood drops. The tongue and the muscles around the roof of your mouth (the palate) relax to such an extent that they block the airways. Also typical is the 'elbow sign': the partner of the snorer regularly pokes this person in the side with their elbow to get the person to stop snoring or to get them to start breathing again. Those with OSAS pay a price: the quality of their sleep drops dramatically, and this can contribute to obesity. People with OSAS are often also tired during the day, occasionally fall asleep, and find it hard to concentrate. Moreover, if you have obesity and also suffer from OSAS, you are at greater risk of developing cardiovascular disease. Although it's not yet entirely clear how these are related, it might be due to the periods of oxygen shortage in the blood. OSAS occurs frequently in middle-aged men with obesity. Some well-known historical figures, such as Napoleon, also suffered from this. So did Winston Churchill, who was very overweight, even though he did write his speeches standing up . . . Luckily, there are ways to treat OSAS. These include a special appliance that brings the lower jaw forward, or, in more severe cases, a special oxygen mask that can be worn at night. This results in better sleep quality and less tiredness during the day. And this can, in turn, provide the energy to start getting more exercise! Because OSAS can be an underlying factor if you are overweight, it might prevent weight loss. So it's important to keep this in mind if you potentially have this and would like to lose weight.

What happens to our biorhythm when we skip breakfast? Because our body's biorhythm is so strong, it's not only what and how much you eat that matters, but also what time it is when you eat it. You've probably already heard that it's not good to skip breakfast. What's more, breakfast is said to be the most important meal of the day. There are indeed studies that show skipping breakfast is linked to weight gain and can contribute to developing obesity, type 2 diabetes and cardiovascular disease. So, seen within this context, it's interesting when you realize that around 20 to 30 per cent of Americans don't eat breakfast, and that, as the obesity epidemic has worsened, breakfast consumption has decreased. Although it is sometimes suggested that it's harder to lose weight if you skip breakfast, there are as of yet no good long-term studies that show that eating breakfast makes it easier to lose weight. A recent study has also shown that by no means all people who skip breakfast gain weight.

But what makes skipping breakfast such a bad idea? You might think you could then 'save' calories. For a long time it was assumed that not having breakfast would lead to increased snacking in the hours that followed. The calories in these snacks would then more than make up for the calories you didn't eat at breakfast, which would result in weight gain. But Professor Daniela Jakubowicz and her research team at Tel

Aviv University discovered that this was not how it worked. What they discovered was that having breakfast affects the small clocks in our organs that regulate blood sugar and insulin levels after meals. They showed that eating breakfast is an effective trigger for setting the clock genes in motion, which keeps the glucose level, blood pressure and weight well under control. This research even suggested that you should have breakfast before 9:30 a.m. to 'set' your clock genes properly. The clock genes in both healthy individuals and those with type 2 diabetes benefited from eating breakfast, and the clock genes of people who skipped breakfast turned out to be less active. So, if you want to maintain your natural biorhythm, start your day with a healthy breakfast and eat it on time!

There are studies that show it's clearly better to eat most of your food in the morning. Of course, this is less about the number of calories in the food than it is about its nutritional value. Researchers at Leiden University Medical Center made an interesting discovery – they found that brown fat in mice is more active in the morning, which means that metabolism is 'turned up' higher. If this also applies to humans, you can 'burn' your meal more quickly in the morning. This once again shows that the old adage that we should 'breakfast like a king, lunch like a prince, and dine like a pauper' still applies in the twenty-first century!

METABOLISM AND THE YO-YO EFFECT

Appetite hormones are easily disrupted, and not just by lack of sleep. Take the yo-yo effect, that familiar cycle of losing weight and then gaining it back. This is a rhythm that no one who is watching their weight wants to experience. What causes the yo-yo effect? In Chapters 3 and 5, we showed that there are biological mechanisms that explain the yo-yo effect, and we also shed light on the fat hormone leptin. But there are even more explanations for this dreaded phenomenon, and they have everything to do with appetite hormones.

THE YO-YO EFFECT FOLLOWING A CRASH DIET: CHANTAL'S STORY

Take Chantal, a personable, spontaneous woman of thirty-four. She had put on more and more weight over recent years. 'As an adolescent, my weight had been relatively normal. Maybe I was a little plump, but nothing worth mentioning. During my first pregnancy, when I was twenty-seven, I gained a lot of weight. While I was pregnant I gained 21 kilos, and afterwards I was only able to lose about half of that. I'd consciously decided to breastfeed my son, because I knew that, in addition to the beneficial effects it would have for him, it would also immediately help me burn extra calories. Sadly, I had to stop breastfeeding after only six weeks due to mastitis.'

During her second pregnancy, three years later, the same thing happened. This time she gained 'just' 18 kilos, but afterwards she was only able to lose about 10 kilos of this. Her husband referred to her affectionately as his 'sweet little walrus'. 'I would try to smile, but deep inside those words cut me like a knife. When I looked in the mirror, I was perfectly aware why he called me that. I did look like a walrus. And an ugly one at that, with a head that was small in relation to my body, and with my groin area and thighs largely hidden behind thick rolls of fat. I was miserable.'

In the years that followed, Chantal gradually gained more weight. She had trouble sleeping through the night and ate irregularly, often skipping breakfast and eating lots of unhealthy snacks between meals. On top of this, she was feeling stressed by the combination of caring for her young children and the demands of her job at the advertising agency where she worked. All of these factors contribute

FAT: THE SECRET ORGAN

to weight gain: hormonal changes during and after pregnancy, lack of sleep, disrupted eating patterns and stress.

Rather than seeking professional help, Chantal decided to go on a crash diet. First she went on a two-week juice fast, and then she moved on to meal replacement shakes that she ordered online. Every day she ate around 500 to 600 kilocalories. 'I denied myself everything. I preferred to skip birthday parties, and I stopped going out to dinner with old friends from school. At the beach, if I was offered ice cream, I'd turn it down. Then my slender girlfriend would push, saying, "Come on, it's just one ice cream!" But I would stand my ground. And . . . it worked! At any rate, I lost around 5 kilos in eight weeks. But I had less energy than before, and didn't carry through on my resolution to do more exercise. While I did take a short walk every evening, how I felt was quite a bit different from what I'd seen on all the various websites, which had promised I would be "bursting with new energy". After ten weeks, I switched to a diet that was as "normal" as possible. I ate three meals a day, and, whenever possible, I ate only healthy between-meal snacks. But I was still hungry, and gradually the number of these healthy snacks – and also the unhealthy ones – increased. No matter how hard I tried, I noticed that my weight was creeping back up, and within a few months I was back at my old weight! I was even a kilo heavier than I had been! How was that possible?!'

Weight gain after being on a strict diet is a familiar phenomenon. In addition to being very frustrating, it often leads to a cycle of rapid weight loss and gain: the yo-yo effect. If they do not go hand in hand with diet and lifestyle changes, strict diets can even be considered to contribute to obesity! This is especially true if this does not involve vigorous exercise. Extremely low-calorie diets (for example, those below 800 kilocalories a day) can only lead to sustained weight loss if they are followed within the context of a complete lifestyle change, especially if this is done with proper guidance. However, the results over the longer term (many years) for those people who both followed a strict diet and received counselling that focused on behaviour change were virtually the same as the results for people who'd only had a behaviour-change intervention. For that matter, strict diets are usually short-lived, which is logical seeing that few people would be able to live on 500 kilocalories a day for the rest of their lives and they would be greatly at risk of developing all sorts of nutritional deficiencies.

The idea that eating (too) few calories over an extended period is good for you comes from animal studies in particular. Professor Jan Hoeijmakers from Erasmus University Medical Center in the Netherlands, who is renowned for his work on ageing, has shown that this is beneficial to survival. When fruit flies, worms, rodents, cows and dogs are put on a diet for their entire lives (known as 'caloric restriction'), the same thing emerges time and again: less food makes metabolism more efficient (read: slower), and less damage is caused in the body at the cellular level, resulting in a longer life. It remains to be seen whether lifelong caloric restriction will also lead to longer lives for humans. In any case, this certainly wouldn't be easy in a world in which people are frequently exposed to high-calorie foods.

And now back to Chantal, who also subjected herself to periods of caloric restriction. How did things work out for her? Once again, she gave it her all, ate extremely little for weeks at a time, lost weight at first, but, in the months that followed, the kilos piled on again in no time. A classic example of the yo-yo effect. Just a few years back, a group of Australian researchers at the University of Melbourne in Australia conducted a ground-breaking study that largely unravelled this mysterious and above all frustrating phenomenon. During the study, fifty people with overweight or obesity went on an extremely low-calorie diet for ten weeks. The levels of appetite-regulating hormones in the blood of the study participants and also their appetite levels were measured before the start of the diet, after ten weeks (the end of the diet) and after more than a year (sixty-two weeks). As expected, the average weight loss of more than 13 kilos was accompanied by a drop in the hormones leptin and insulin. The hunger hormone ghrelin actually increased. Other appetite signals also fell into line with this, and what it came down to was that the levels of the hunger hormones rose and those of the satiety hormones dropped. So the study participants should have felt hungrier. And, sure enough, the participants indicated that their appetite had increased after the crash diet.

But now comes the ground-breaking part. A year after the initial weight loss, these hormonal changes were still present, and resulted in increased appetite and a less pronounced feeling of satiety! And these people also still experienced more intense feelings of hunger. These findings shocked the scientific world and led to many new insights into the yo-yo effect. Up until then, it had always been assumed that, once you went off a crash diet, your body would quickly recover. But this study showed for the first time that crash diets can interfere with your hunger and satiety system for a long time. And it's not clear if this system will ever recover, because longer-term data are not yet available. What we do know is that, when you're on a crash diet, your metabolism slows down, so not only does your appetite increase, this also means you can't eat as much without gaining weight. A study of participants in the American weight-loss race *The Biggest Loser* showed just how far this goes. During this race, people lost weight using a combination of strict dieting, vigorous exercise and coaching. Over the course of thirty weeks, they lost on average a whopping 58 kilos of body weight. But they paid a high price for this, because at the end of the race their metabolic rate had dropped sharply – by 600 kilocalories a day – even though they had started exercising more. Six years later, the participants had gained back 41 kilos on average, and their metabolic rate remained at the same lower level. A shocking conclusion that can be drawn from this is that, if a person who has ever had obesity loses weight by going on a strict diet, their metabolism will be slower and they will ultimately not be able to eat as much as someone of the same weight who has never had obesity. Even if they've lost weight, many factors in their hormonal system and metabolism will still be disrupted, and their body will be programmed, as it were, to easily gain weight again. For some people, then, a crash diet can be an excellent recipe for weight gain.

Scientists are currently concentrating on coming up with strategies that prevent our hormonal system from becoming disrupted by a low-calorie diet, and that leave metabolism intact. Only then can we ensure that fanatical dieters who do their utmost to lose weight will achieve real success in the long term. Right now, those who do their very best and deny themselves everything (and don't forget the social function that food plays!) are hit the hardest. And not only because others may perceive them as being 'weak' if they go off their diet and gain weight again. They often judge themselves harshly, and feel frustrated and ashamed that they've failed yet again, without realizing that the biology of the hormonal appetite regulation system makes it very easy for them to gain back weight after a crash diet.

Sometimes juice fasts and other crash diets are used to cleanse the body; this is also referred to as 'detoxing'. The idea behind this is that this would give your digestive system a rest and allow your organs to focus on getting rid of any waste products. This is complete nonsense. The body already has an excellent system for cleansing itself, namely, the gastrointestinal tract, in which intestinal cells are regularly replenished by stem cells and the gut bacteria that are present keep an eye on things. By far the best way of maintaining the gastrointestinal tract is with healthy food that contains enough fibre and nutrients, and also enough fluids (such as water). And of course our kidneys and liver expertly filter out any toxins that our body may contain. So don't be fooled into buying such senseless, overpriced 'detox' or 'cleansing' products. Instead, put the money into your children's piggy bank or buy a nice gift for your neighbours – this is something that will actually contribute to your wellbeing!

SOME DIETING TRENDS AT A GLANCE

In addition to the crash diet, people have come up with all sorts of ways to lose weight and keep it off, or to maintain a healthy weight and avoid the yo-yo effect. For instance, there are those who swear by being on a lifelong 'diet'. These people refer to themselves as 'CRONies' (CRON stands for 'Calorie Restriction with Optimal Nutrition'). They consume around 1,800 kilocalories a day, which is 10 to 30 per cent less than what they need. These people cut back mainly on the amount of protein, and make sure they get enough fibre, vitamins and minerals. According to Hoeijmakers' concept, eating less over an extended period puts the body into energy-saving mode, and because less growth takes place there is less cell division and probably also less damage to the DNA. The body will then automatically take better care of the cells, and this delays ageing. The initial results from people who have been on a CRON diet for an extended period of time (fifteen years on average, so not yet for their entire lives) do indeed show beneficial effects on overall health – but this does mean always having to live on restricted rations.

Others experiment with their eating rhythms. At the moment, 'intermittent fasting' is very hip. This works by restricting calories for a certain period of time. For example, one day you eat less than 25 per cent of your energy requirements, and the next day you can eat whatever you like. A variation of this is 'periodic fasting', when you fast for one or two days a week (and five or six days a week you don't). The idea behind intermittent fasting is that you retain the advantages of calorie restriction without slowing down metabolism or disrupting the 'hunger hormone system'. Because there are all sorts of different intermittent fasting schedules, this makes it hard to conduct large studies and to interpret them accurately. Through animal studies, researchers have discovered the beneficial effects intermittent fasting has on the brain, intestinal flora and ageing, and have also observed a drop in the number of inflammation markers. Most studies in humans have only shown minimal weight loss and minor effects on metabolic parameters like glucose and cholesterol levels. The effects of intermittent fasting on weight reduction are comparable to those of calorie restriction, and so in the short term this seems quite favourable. So is this the holy grail of weight loss? Unfortunately, not enough is currently known about the long-term effects, so this remains to be seen.

Another interesting phenomenon is 'time-restricted fasting'. Although this is still in an experimental phase, this method appears very promising. Here, there are no restrictions on the number of calories a person is allowed to eat, but the time span during which someone eats over the course of a day is drastically reduced. So, a person can continue eating about the same as they did before, but they do this within a shorter period of time. In most studies, this period was six to a maximum of twelve hours. Outside of this time frame, the person doesn't eat, but is allowed to drink water. Studies in rodents have shown that restricting the time frame during which eating takes place has beneficial effects on weight and metabolism. There is initial evidence that this also applies to humans, but more research is needed to be able to see how great these effects really are. The American researchers Shubhroz Gill and Satchidananda Panda published an interesting study in the prestigious journal *Cell Metabolism*. They used a smartphone app to study the eating patterns of adults who did not work night shifts. Most of them ate very frequently, and what was striking was that they ate the fewest calories (< 25 per cent) in the morning, and ate the most calories (> 35 per cent) in the evening, after six p.m. This means that the time during which they ate regularly was spread out over nearly fifteen hours on average. It turned out that, when those who were overweight and used to eating during a fourteen-hour period every day reduced this to ten or eleven hours, they lost weight, had more energy and slept better! And these effects were still present a year later. These effects also appear to be related to our biological clock. It thus appears that we need to pay close attention not only to what we eat, but especially to when and within what time frame we eat!

THE MENSTRUAL CYCLE AND FOOD CRAVINGS

And finally, a very different biorhythm is a woman's menstrual cycle, which fluctuates not over the course of the day, but over the month. One or two weeks before they get their menstrual period, some women suffer from mood swings: they feel blue, are irritable and crave high-calorie foods like chocolate. These characteristic cravings disappear again once their period is over. Although this scenario might seem familiar to many women, we don't yet know exactly what causes this. Just being aware of these cravings can already help, so that they don't expose themselves to the things they crave (for example, by not having these things in the house).

All kinds of different rhythms are at work in our body without us realizing it. Often, we only notice this when we disrupt these finely tuned natural rhythms through lack of sleep, by travelling through different time zones, by eating at the wrong times (like the middle of the night!), or by going on crash diets. And all of these disruptions can contribute to weight gain. There is an old Dutch saying that says our lives should be governed by 'rest, hygiene and routine', and this nugget of wisdom still applies today.

8

How does stress cause overweight?

Imagine the following situation. You're in the cabin of a small aeroplane nearly four kilometres above the ground, and it's cramped and stuffy. You look out and all you can see is farmland; by now, the houses have become tiny specks. Your breath catches in your throat from fear, and your heart is hammering in your chest. A multitude of thoughts race through your mind: why did I want to go skydiving so badly? What were the landing instructions again? Pull down the lines and put your legs out in front of you. Right? What if the parachute doesn't open? What if . . . if . . . ? The door of the plane opens and, calmly but firmly, the instructor says, 'We're at the right altitude – we can jump now. Put your head back and let yourself fall . . . go!'

What you're experiencing right now is stress – acute stress. What is happening in your body at this moment is a beautiful biological interplay. Your brain is sending signals by way of nerve fibres telling your adrenal glands to make substances like adrenaline. This happens in a flash, and immediately sets a process in motion in your hypothalamus (the 'air traffic controller') that releases substances that boost stress hormones. These substances send a signal to the pituitary gland, the large 'master gland' that is located behind the bridge of your nose and below your hypothalamus. The pituitary gland in turn sends a different kind of regulating hormone (ACTH), which travels through the blood-stream to both adrenal glands. During this acute state of stress, the adrenal glands then pump lots more of the stress hormone cortisol into your blood.

The stress hormones adrenaline and cortisol cause your heart to beat faster and your blood pressure to immediately shoot up. This is helpful, because more blood will then be pumped through your body. This is necessary for transporting extra sugars and oxygen to places like your brain, which will allow you to think faster and more clearly. Ideal during an exam. So a small case of the nerves is actually good for you. More energy in the form of sugars is also transported to your muscles. These sugars can be released very quickly from the glycogen stores (which we mentioned earlier) in your liver and muscles. This is also something these stress hormones are responsible for. Handy if you are suddenly charged by a hungry tiger and you need to run away fast . . . Your muscles can rapidly convert these sugars into energy and movement – although, when you're four kilometres off the ground, tigers would be the last thing you'd be worried about.

In the end, the entire chain reaction of biological stress hormones comes to a halt because cortisol slows its own production via both the pituitary gland and the hypothalamus, and as a result the body's stress response is extinguished. And, if all goes well, you land safely with both feet on the ground.

Stress is currently a hot topic. According to the American Psychological Association, the majority of Americans feel moderately to very stressed. This is also true for many people worldwide: hard-working managers or factory workers, overworked teachers, students of all ages who suffer from performance stress and also have busy social lives (both off- and online), single mothers trying to keep their heads above water.

By stress, we usually mean an imbalance between what a person can handle and what is asked of them or what they expect of themselves. And this imbalance often arises due to a combination of stress factors at work and in their personal lives. Stress affects the brain in all kinds of ways. For example, a short period of intense stress can enable us to perform certain tasks exceptionally well. This can be useful during a sporting event, an exam or for meeting an important deadline. But if the stress is too intense, our performance usually actually goes down. If stress factors persist and there is a lack of support, mental stress can become a chronic problem.

What is striking is that this 'stress epidemic' came into being at approximately the same moment as the obesity epidemic. Right now, worldwide, 39 per cent of all adults are overweight. There are more and more indications that these two epidemics are related, and science is providing more and more evidence that stress can make you fat.

MENTAL STRESS, PHYSICAL STRESS AND WEIGHT GAIN

Before we can address the question of why stress and overweight often go hand in hand, it's important to realize that there are different kinds of stress. In addition to acute mental stress (such as during a parachute jump), there are also different kinds of physical stress.

We've all been through the following: crawling into bed and spending one or more days shivering with fever while your head feels like it's stuffed with a big wad of cotton wool. You have no appetite, and your muscles feel weak, even painful. In short, you've come down with a bad case of the flu. This, too, is a form of stress, and the body makes extra cortisol in response. When the stress is caused by illness, the body doesn't respond to signals from the brain that register stressful events, but to inflammatory substances that are released when fighting the flu. These substances tell the brain there is a viral infection present, and that the brain should send signals to reduce the inflammation. Then the hypothalamus and pituitary gland make extra regulating hormones that prompt the adrenal glands to produce extra cortisol, because one of the functions of cortisol is to lower inflammation as part of the recovery process.

Precisely because of their anti-inflammatory effect, drugs that contain cortisol-like substances – known as corticosteroids – are frequently used for inflammation, inflammatory diseases, and conditions in which the body's own immune system is overactive, such as asthma or rheumatoid arthritis. So, in fact, inflammation is also experienced by the body as stress due to the increased cortisol production. The same goes for chronic pain, because this results in sustained activation of the stress system. In the previous chapter, we saw how lack of sleep and disruption of our day/night rhythm can also lead to a stress response in our body. In short, stress is a broad concept in which the body makes little distinction between psychological and physical stress. In both forms of stress, the body goes into survival mode and starts producing more cortisol.

AN EXAMPLE OF THE CONSEQUENCES OF EXTREME STRESS: MILA'S STORY

Mila is a forty-one-year-old primary school teacher; she is married to Jacob and is the mother of three children. She goes to the gym on a regular basis, and cycles to work every day. Over the years, though, she started to feel that something in her body was changing. 'At the gym, I was no longer able to push away the weights with my legs. My muscles got more and more weak. I gained weight, and started getting these wide purplish-red stretch marks on my belly. My face was puffy, and I had red cheeks – I didn't need to wear blusher any more. I went from size 38 to size 42, and my periods became more irregular.'

Mila's memory also deteriorated rapidly. She forgot the names of the children in her class, and, much to her dismay, she could no longer remember what she'd talked about during parent – teacher conferences. She noticed that her husband was distancing himself from her. 'When I looked at myself in the mirror, I saw someone else – a woman with a round belly, a bulge of fat on the back of her neck and an unhappy look in her eyes. I developed facial hair, while the hair on my head just kept falling out. My sex drive had frozen solid. I was disgusted with myself . . . One day, I was waiting in the schoolyard to pick up my child. I was wearing my favourite black flowered top, and apparently my belly was clearly visible beneath my shirt. A child who had just come out asked cheerfully, "Are you pregnant?" I wanted to sink into the ground.'

Then Jacob dropped a bombshell: he no longer found Mila attractive. He thought she'd allowed herself to become a fat, bad-tempered person, and told her he was in love with another woman. This hit Mila hard.

A couple of months later, Mila was hospitalized with a severe intestinal inflammation. On a scan, the medical staff happened to see that her right adrenal gland was enlarged. Further tests in a number of different hospitals showed that a lump in this adrenal gland was producing too much cortisol, round the clock. So, stress hormones were constantly raging through Mila's body, causing a wide range of symptoms. The medical diagnosis came back: she had Cushing's syndrome. 'So that explained my weak muscles, my big belly and why my brain wasn't working the way it should!'

Mila had surgery on her adrenal gland, and the hormone-producing lump was removed. In the months that followed, Jacob realized that Mila was ill and therefore not herself, and so she could do little about the changes to her appearance and personality. His love for her was rekindled, and he decided to move back in with her and the children.

Unfortunately, this story doesn't end with 'and they lived happily ever after'. During the year after her surgery, Mila began to have panic attacks, still had a large belly and memory problems, and developed muscle pain and extreme fatigue. She did her utmost to get back to work again during the rehab phase, and the medical professionals treating her prescribed medication. Even so, things did not go well enough, and she lost her job.

A number of years have passed since then, and things are slowly improving. The puffiness in her face has disappeared, her energy is returning and her fears are diminishing. She goes to the gym again, and is gradually losing some weight.

Mila's story illustrates what can happen if you have extremely high levels of cortisol in your body for an extended period of time. We know that chronic stress – whether its source is psychological or physical – can result in health problems. Cushing's syndrome – the lump in Mila's adrenal gland – is very rare. But you could see it as a kind of 'model' of an extreme form of chronic stress that causes the body to produce far too much cortisol. This condition is caused by physical rather than mental stress, and shows the profound effect excess cortisol has on the body, including on body fat. Within a short period of time, a person's belly fat increases, they develop a bulge of fat on the back of the neck, and their face gets rounder, while the amount of subcutaneous fat on their arms and legs actually decreases. Their muscle mass also decreases, which means they have less strength in their arms and legs. Their blood pressure also goes up, their cholesterol and sugar metabolism becomes disrupted, and they may start to feel down. In addition, there are many other typical symptoms associated with this rare syndrome, such as wide purplish-red stretch marks (striae), acne, thin and very delicate skin, spontaneous bruising, poor wound healing and, in women, a disrupted menstrual cycle and excessive hair growth. Some women suddenly find they have to shave their face on a daily basis!

If the lump causing the cortisol overproduction is surgically removed, many symptoms disappear. However, people often continue to have memory problems and excess belly fat for a long time.

WHY DOESN'T EVERYONE
BECOME OVERWEIGHT FROM
CHRONIC EXTREME STRESS?

People differ considerably in the way their bodies deal with stress. To start with, every person experiences stress differently. Any stressful event – such as the death of a loved one – can be experienced by individuals in different ways. One person might grieve deeply for months over the loss of a pet, experience sleeping problems and heart palpitations, while, after the loss of a partner, another person might bounce back quickly and be able to resume their regular routine without experiencing any physical symptoms.

To understand why a continuous stress signal affects one person more than another, it's good to know that the stress hormone cortisol can't do its job on its own. Just like every other hormone, cortisol needs a receptor (the 'hormone receiver') to transmit this signal to the body cells. This is known as a corticosteroid receptor, and is found in virtually every kind of cell in our body, including in our body fat. It's interesting how different people are in terms of their sensitivity to cortisol, and this difference is determined mainly by the sensitivity of the corticosteroid receptors. This sensitivity is largely hereditary, and thus already present at birth. An important role is played in this by the gene responsible for the corticosteroid receptor: the corticosteroid receptor gene. Variation in the DNA codes of this gene mean that one person will be more sensitive to cortisol than another, just like our DNA determines whether a person will have brown, green or blue eyes.

Nearly half of people are carriers of a specific genetic variant of the corticosteroid receptor gene, and Liesbeth van Rossum's research team at Erasmus University Medical Center in Rotterdam discovered that this variant is associated with an increased sensitivity to cortisol. Another interesting finding was that carriers of this cortisol-sensitive variant more frequently have a larger belly, poorer cholesterol and sugar metabolism, less muscle mass and a greater risk of depression: all characteristics we also see in long-term overexposure to cortisol, such as with Cushing's syndrome, which is what Mila had.

In contrast, there is also a genetic variant of the corticosteroid receptor which occurs in 5 to 10 per cent of the population that was found to be associated with a relative insensitivity to cortisol. And there we see the opposite – and actually beneficial – effect on health. For example, we observed that men who were carriers of this variant had more muscle mass and muscle strength, and were taller. Female carriers of this cortisol-insensitive genetic variant tended to have a smaller waist, which indicates less belly fat. In both male and female carriers, there was an association with a beneficial metabolic profile – that is, a lower risk of diabetes and lower cholesterol levels. It also emerged that carriers of the cortisol-insensitive variant generally lived longer. It would seem that this small group had been made 'biologically stress-resistant' at birth. In this way, two people who are the same age and have comparable diets and lifestyles, and experience the same amount of stress, can be very different in terms of their fat because one of them has been 'blessed' with a cortisol-resistant genetic variant and the other has not.

CAN YOU MEASURE STRESS?

Before research can be conducted on whether stress can make you fat, we first have to be able to measure stress. Although there are questionnaires for mental stress that indicate how much stress a person experiences, of course they don't measure how your body's stress system responds to this from within. To do this, the cortisol in the body would have to be measured, and one way this can be done is in the blood. But, as we saw in the last chapter, there's a problem with this: your cortisol level does not stay the same over the course of the day.

Cortisol has a day/night rhythm, rising during the final hours of the night and peaking just before we wake up. Moreover, many people find having blood drawn in itself a stressful event, and, as a result, the measured blood level may primarily reflect how afraid someone is of needles . . . Cortisol can also be measured in saliva or urine, but these methods also have their limitations, and both of them tend to say more about the stress level of that particular moment than about chronic stress.

As researchers, what we would really like to measure is a person's cortisol level over time. To get an impression of this, we introduced a method that was already common in forensic medicine. It turns out that cortisol is easy to measure in the hair on our head. Cortisol is absorbed into our scalp hair from our bloodstream. Because our hair grows an average of 1 centimetre a month, every centimetre of hair provides an average of the cortisol levels in the body during that month. So the 3 centimetres closest to the scalp reflect the last three months. With longer hair, hair analysis even offers the opportunity to construct a historical timeline of an individual's biological stress level, which can be compared to the annual rings of a tree.

Scalp hair makes it possible for researchers to study whether people with obesity had higher long-term cortisol levels, without needing to use needles or take urine samples. At Erasmus University Medical Center in Rotterdam, we used this Crime Scene Investigation-like method to study the relationship between stress and obesity. And it worked! We were able to conclude that the cortisol levels in the hair of adults with obesity were indeed much higher than those of people with a normal weight. We then wanted to know if this was also true for children. To do this, we studied more than three thousand six-year-olds in Rotterdam, and found that the children with the highest hair cortisol levels were nearly ten times as likely to have obesity. Another study we conducted among more than 280 elderly people showed that those with the most cortisol were around two and a half times as likely to suffer from cardiovascular disease. In this group, in terms of risk, this was of the same magnitude as the association between smoking, high blood pressure, and diabetes on the one hand and cardio-vascular disease on the other hand. Perhaps in future, a visit to the general practitioner will include not only an assessment of these classical risk factors for cardiovascular disease, but also the biological stress level.

However, further research is still needed. Because, is this association causal? In other words, does more cortisol lead to obesity, or does obesity lead to more cortisol, and is this why we measured higher cortisol levels in people with obesity? We already knew from Cushing's syndrome that too much cortisol can lead to weight gain. But there are more and more strong indications that obesity can, in turn, lead to a high cortisol level. Research on this is still underway.

HOW DOES CORTISOL MAKE YOU FAT?

One of the most noticeable effects of prolonged high cortisol is that it leads to redistribution of our body fat. This happens when cortisol binds to its corticosteroid receptors in the fat cells. This results in a decrease in fat mass (and also muscle mass) in the arms and legs, while fat in the abdominal area actually increases. As we now know, abdominal (belly) fat is the most harmful kind of fat. When body fat increases, it can produce certain kinds of fat hormones and inflammatory substances that can contribute to developing diabetes, atherosclerosis and unhealthy metabolism, and can even have a negative effect on mood.

A regular joke among obesity and stress researchers is that 'stressed spelled backwards is desserts'. And whether this is a coincidence or not, these two words are indeed closely linked. In other words, stress makes you crave high-calorie foods – a second important mechanism through which a high cortisol level can lead to higher body weight. Cortisol sends signals to the appetite regulation centre in the hypothalamus in the brain, and you start to feel hungry – particularly for foods high in fat and sugar. In other words, you crave snack foods! This might explain why, during a stressful time, you're more likely to go for a big bar of chocolate than a salad. And, to make matters worse, there is evidence that suggests that things like sugary foods can lead to an increase in cortisol – a vicious circle!

Cortisol can contribute to the development of obesity in various ways. For example, it could start with someone eating too many unhealthy foods over an extended period of time, causing them to gain weight and eventually become obese. Then, the increased amount of belly fat this person has developed can cause their cortisol level to go up in a number of ways. Which makes them crave snack foods even more. And today these unhealthy snacks are sold on every street corner . . . In the Western world, we are exposed daily to unhealthy foods, and more frequently than we are to healthy ones. And what happens? You eat even more, which then makes it much more difficult to lose weight.

But a person might also find themselves in a stressful situation for an extended period of time. Maybe you have financial problems, or you've just lost your job, or you're going through a divorce – this mental stress boosts cortisol production even more. The resulting elevated cortisol level can lead to cravings for high-calorie foods, which in turn leads to weight gain – and more belly fat in particular, which further increases cortisol levels. Also here, people can end up in a vicious circle. This is why just giving someone dietary advice ('Eat lots of carrots, and good luck') is usually not enough if the underlying stress isn't dealt with. This requires an approach that also focuses on the psychological component: mental stress, disrupted sleeping patterns, chronic pain and other stress factors, as well as our (eating and exercise) behaviours and our habits.

Another way that cortisol might contribute to weight gain is through its effect on our brown fat, the 'good fat' that converts calories to heat. The more active your brown fat is, the better – at least, if you want to lose weight or maintain a healthy weight. Animal studies have shown that a high cortisol level may suppress the activity of brown fat. Although this may also be true for humans, as we mentioned earlier, a mouse is not a human. Further research is needed to provide a definitive answer on this.

Alcohol can also increase your cortisol level. Many people regularly pour themselves a glass of wine, or have a beer or something stronger, when they're under pressure. Before a short stressful event, such as standing in front of an audience, an alcoholic beverage does indeed inhibit the acute stress response and makes a person a bit more relaxed. But this is certainly not to be recommended!

Chronic drinking is another story entirely. Excessive alcohol consumption occurs at all levels of society, and goes beyond the clichéd images of potbellied men drinking beer and watching football on the couch, or homemakers who are overly fond of sherry. Alcohol consumption also tends to be high when people get together after work on Friday for drinks, at student associations and in the canteens of sports clubs.

Scalp hair analysis has shown that the cortisol levels of people who are chronic excessive drinkers are three to four times higher than those of people who don't drink alcohol or who used to have an alcohol problem but no longer drink. A beer belly can be put down in part to an elevated cortisol level. In the medical world, the clinical picture of a person who is consuming excessive alcohol is sometimes referred to as pseudo-Cushing's. Along with more belly fat because of an elevated cortisol level, excessive alcohol consumption can also contribute to weight gain by causing a slight drop in testosterone (in men), by suppressing fat burning, by inhibiting our impulse control (which means we're more inclined to grab a snack), and of course by contributing to our total caloric intake without inducing a feeling of satiety. You could see alcohol as a hidden contributor to overweight. And it's something that many people get too much of. According to the WHO, harmful use of alcohol is responsible for 5.1 per cent of the global burden of disease. According to the Centers for Disease Control and Prevention (CDC) in 2013, more than half of the US adult population had consumed alcohol in the thirty days prior to the survey. About 6 per cent of the adult population reported heavy drinking, and 17 per cent reported binge drinking.

If you happen to be reading this chapter while having a glass of wine on the sofa, no worries. But if that glass of wine is a daily habit and if you'd actually like to lose a few kilos, consider having a cup of tea or coffee next time you sit down to read. In the next chapter, we'll talk more about hidden contributors to overweight that aren't nearly as pleasant as that glass of beer or wine.

9

Hidden contributors to overweight

INJECTABLE STRESS: JULIE'S STORY

By now, it should be clear that stress has a tremendous impact on our body. And sometimes this stress comes from unexpected sources. This was the case for twenty-four-year-old Julie, an avid athlete who visited the Obesity Center CGG at the Erasmus University Medical Center in Rotterdam one day. She was clearly having a hard time. In the consultation room, she seemed timid and vulnerable, and looked down at the floor. She said her weight had always been normal. She'd always been in good shape, and if she could she would run or play basketball every day. This had all changed abruptly six months earlier, when, for no apparent reason, she'd gained 14 kilos. She had developed a belly and her cheeks were full and flushed. She said that her diet hadn't changed, and she was still eating lots of fresh fruits and vegetables. Then she broke down. In tears, she went on with her story: 'I don't know what's the matter with me. I'm studying to be a lifestyle coach, and I'm doing an internship in a practice where I support people with overweight in developing a healthy lifestyle. I feel like no one takes me seriously any more because I've been gaining so much weight myself within a short period of time. It feels like everyone is giving me disapproving looks – the hardest are the ones that drift down to my belly, which just keeps getting larger. I feel so insecure – I really feel like I've failed.'

During her visit, we thoroughly discussed all of the factors that might have contributed to her weight gain. Her diet wasn't the problem, nor was her active lifestyle. In the months before she gained the weight, though, she'd had a lot of problems with her knees, probably as a result of all the running she did, and she'd been given a number of injections. Now that she thought about it – yes, the problems had started after she'd had the injections.

At the Center, we wanted to know if there might have been corticosteroids in the injections. This kind of drug is a 'sister' of the stress hormone cortisol, and one of its uses is to treat joint inflammation. And, sure enough, this 'injectable stress elixir' turned out to be to blame for Julie's weight gain. The corticosteroids had ended up not only in her painful joints, but also in her bloodstream. There, just like her body's own cortisol, they were able to exert their effect via the corticosteroid receptors. As a result, Julie gained weight, especially in her belly. This corresponded with the story of Mila, who had Cushing's syndrome, and whose excess cortisol had been produced by her own body.

Injecting a synthetic stress-hormone-like substance can have far greater consequences than many people think. In more than half of those who are given a corticosteroid injection to treat joint inflammation, the adrenal glands start to make less cortisol and as a result become 'lazy'. So the corticosteroid drug that entered the bloodstream has an effect on the adrenal glands by way of the brain.

There are corticosteroids in many medications. The most well-known of these are prednisone and dexamethasone tablets, but there are also many drugs containing corticosteroids that are administered locally, either on or in the body. These include inhalers for asthma, nose sprays for allergies, eye drops, ear drops, enemas for the bowel and creams for skin conditions like eczema. Fortunately, with these local medications there is a much smaller risk of developing a lazy adrenal gland, but if people are using multiple corticosteroids, this risk increases sharply. A combination of asthma, eczema and hay fever, with use of locally applied corticosteroids for each condition, is not unusual. On the contrary – these immune system disorders often go hand in hand.

Corticosteroids are prescribed frequently in many countries worldwide. These drugs are extremely effective for treating a wide range of conditions, and are sometimes very necessary. Our research showed that, if you were to take a cross-section of the Dutch population, around 10 per cent use corticosteroids. More than a quarter of people living with obesity use some form of corticosteroids. In the Netherlands, corticosteroid-containing products are only available on prescription, and not over-the-counter, in contrast to many other countries, like the US, China, Africa and India. So, the high Dutch numbers may even be an underestimation of the global use. The most-used types of corticosteroids are inhalers, ointments, nose sprays, pills and injections that contain corticosteroids. They might even be in your own medicine cabinet.

CHEMICAL STRESS

It is generally known that tablets containing corticosteroids, like prednisone and dexamethasone, lead to weight gain, especially in the abdominal region. These drugs also make you much hungrier. Although these kinds of medications are very effective for a wide range of conditions, these side effects are one of the reasons not everyone is keen to take them. What is less well known is that corticosteroids administered locally, somewhere in the body, can also lead to weight gain, as we saw in an extreme way with Julie. In many people, the side effects of corticosteroids administered locally are negligible, but, even so, corticosteroids are not entirely harmless. Many of the people with obesity who visit the Obesity Center CGG in Rotterdam turn out to be using corticosteroids. Although treating a patch of eczema with an ointment for a couple of weeks will not immediately lead to weight gain, people who use lots of these medications on a daily basis for an extended period of time could indeed gain weight from this. This called for further investigation.

An initial assessment showed that the percentage of corticosteroid users among people living with obesity was two to three times higher than it was among people with a 'normal' weight. Furthermore, in a larger study of more than 140,000 Dutch adults, we found that corticosteroid users – both those who used the 'big guns', like prednisone, as well as those who used the 'harmless' locally administered medications – had a higher BMI and, more importantly, they had a larger waist circumference. This larger waist circumference in particular is what you can expect from stress hormone cortisol's 'little sister', as we saw from the stories of Julie and Mila.

In short, these studies seem to indicate that local corticosteroids contribute to weight gain, although a causal relationship needs to be proven. This association was particularly apparent for inhalers and nose sprays. Inhalers are used mainly by people with allergic asthma. Many people who are living with extreme obesity and also have lung problems that are presumably related to their obesity are often mistakenly diagnosed with 'allergic asthma'. Research has shown that this was the case in almost half of those diagnosed with both obesity and asthma. When a person is severely overweight, their lung capacity is reduced and there is increased resistance in their respiratory tract. Although this produces symptoms that seem to point to allergic asthma, this is not the case, and so inhaled corticosteroids have little effect! Moreover, in people who have obesity and do have asthma, the fat hormones released by their sick and inflamed fat might have contributed to their shortness of breath. Recent studies have shown that weight loss through intensive lifestyle interventions or bariatric surgery has many beneficial effects on asthma in adults with obesity, and these methods should therefore be part of their treatment for asthma.

In addition to corticosteroids, there are countless other medications that lead to weight gain (see **Box 10**). These include certain antidepressants and antiepileptic drugs. And there are also the notorious drugs used to treat psychosis (known as antipsychotics), such as clozapine, olanzapine and risperidone. In many cases, antipsychotics affect leptin and ghrelin levels, which causes people to get hungrier, eat more and thus gain weight. Because we know more and more about all of the consequences of obesity, weight gain in people with psychiatric conditions is also being taken increasingly seriously.

Box 10. Examples of medications that can have weight gain as a side effect

- **Corticosteroids** (local application, tablets or injections)
- **Medications for treating high blood pressure** (beta blockers, alpha blockers)
- **Antidepressants** (mirtazapine, citalopram, paroxetine, lithium)
- **Antipsychotic drugs** (olanzapine, risperidone, clozapine, quetiapine)
- **Antiepileptic drugs** (carbamazepine, valproic acid, gabapentin)
- **(Neuropathic) painkillers** (pregabalin, amitriptyline)
- **Antidiabetic drugs** (insulin, glimepiride)

A possible association with weight gain has been found for the following:

- **Stomach acid inhibitors** (proton pump inhibitors)
- **Allergy medication** (antihistamines)

Having a psychiatric condition is hard enough, and things just get harder if you also run the risk of developing obesity – as well as the conditions associated with this, such as diabetes and cardiovascular disease – from the medications you need to take. This also results in a lower life expectancy for people who take these drugs. Fortunately there are various ways to limit weight gain when taking antipsychotics. First of all, before people start taking these drugs, it helps to make them aware that they run the risk of gaining weight, so that they can take steps to prevent this: a healthy diet, more exercise and cognitive behavioural therapy can all help. For example, before starting a treatment with antipsychotics, one woman sets out some fruits and raw vegetables so she's prepared for any nocturnal eating binges. Since she started doing this, she has also gained less weight than during previous periods of antipsychotic use, when she would binge on snack foods in the middle of the night. Many people are taking antipsychotics. A study in sixteen countries worldwide, by Óskar Hálfdánarson and colleagues, showed that overall antipsychotic use across all ages was highest in Taiwan (78.2 per 1,000 persons) and in the publicly insured US population (40.0 per 1,000 persons). Because of this, the search is on for medications that can prevent weight gain and that can also be taken in combination with antipsychotics. The most effective of these is metformin, a drug used to treat diabetes that also has a positive effect on weight and that can curb appetite. Some people may still gain a lot of weight in spite of changing their lifestyle. If this happens, their doctor may consider replacing the antipsychotics with medications that lead to less weight gain, but this is often tricky. And part of an individual's susceptibility to weight gain when taking antipsychotics is determined by a person's genes.

What many people don't know, and what is quite shocking, is that you can also gain weight from taking medications used to treat conditions related to obesity. Examples of such drugs are certain beta blockers, which are often used to treat high blood pressure, and insulin, which is used to treat diabetes. So, when people who have obesity adopt a healthy lifestyle and try to lose weight, these drugs interfere with their attempts! It's very sad to see these people struggle. In spite of all the good advice they put into practice, they don't lose a single gram. It's a little like trying to drive a car with the handbrake on, because, as we mentioned earlier, high dosages of insulin actually cause your body to store more fat, and beta blockers lower the resting metabolism. The latter is due in part to the inhibition of brown fat and the lowering of the heart rate, which can lead to fatigue and make it harder for people to exert themselves and exercise vigorously. Fortunately, your doctor can help to lower the dosages where possible (don't try to do this on your own!), and in some cases it may be possible to stop using a drug altogether. This also makes it much more likely that a healthy lifestyle will 'take'. If a person manages to lose weight, there is also a much greater likelihood that they can stop taking (or reduce) their medication for high blood pressure, diabetes or depression. We're not trying to say that these medications are unnecessary or ineffective. However, it's also good to be aware that they sometimes interfere with weight loss.

HORMONE DISRUPTERS

Everyone knows that, when you take medication, you ingest chemical substances that set something in motion inside your body. Many people will also know that some of these substances can lead to overweight. But did you know that a plastic water bottle is also a source of hidden contributors to overweight? As are the plastic toys your daughter, son, niece, nephew or grandchild plays with. Every day, we unwittingly expose our body to 'hormone disrupters', and these have been linked to negative health effects such as reduced fertility, breast cancer and also obesity.

You can think of a hormone disrupter as a chemical substance that can both mimic and block the functioning of our natural hormones. Common hormone disrupters are bisphenol A (BPA) and phthalates (plasticizers that make plastic items more flexible). Even in extremely low concentrations, these substances are known to have the ability to disrupt our hormonal system, although it isn't clear to what extent. Therefore, scientists and government agencies are currently conducting extensive research into the effects of hormone disrupters.

Hormone disrupters enter our body unnoticed, because they leach out of the materials our food comes packaged in. Examples of this include water bottles, the insides of food tins, and plastic plates and cups that are heated in the microwave. To protect babies, baby bottles containing BPA are now banned in the European market.

Plastic dolls and soft plastic toys, which some little children like to put in their mouth, can also contain hormone-disrupting phthalates, and soft toys can contain hormone-disrupting flame retardants. There are also hormone disrupters in electronics, medical equipment, pesticides, cosmetics and in many more of the products that surround us. They are even found in such everyday items as cash register receipts, sunscreen, shampoo, shower gel, day creams, nail polish and lotions. Hormone disrupters enter the body through the skin or the mouth, or they can be inhaled. So maybe you really can get fat just from breathing . . .

Scientific studies are making it increasingly clear that hormone disrupters can interfere with our energy and fat-cell metabolism. At the same time, we should not yet draw any firm conclusions, or start wearing a protective suit and mask immediately after reading this, because most of these data are from animal studies, and the implications for humans are still far from clear.

Both animal studies and studies in humans have already found associations between BPA and the hormone leptin, which curbs appetite, and the hormone ghrelin, which stimulates appetite. Hormone disrupters also seem to influence the sensitivity to sugar and our fat metabolism, and in this way can lead to weight gain. Exactly how some of these substances enter our body is as surprising as it is shady. It's like a bad thriller, only this time it's not the butler or the ex-mother-in-law who tries to poison the man of the house by adding some chemicals to his food. As mentioned, some hormone disrupters can slip into our body unnoticed via the food we eat because they 'leak' out of the materials they're packaged in. So, are you starting to get a little paranoid? Wait – it gets even more frightening. Humans are probably at their most vulnerable to these substances just before or just after birth. Animal studies have shown that pregnant mice who had been exposed to certain hormone disrupters had offspring who were much heavier than those of the mouse mothers who had not been exposed to these disrupters. Newborn animals who were given the synthetic oestrogen DES (diethylstilboestrol) or BPA also gained weight much more quickly than animals who had not received this.

Although we don't yet know with certainty whether this effect is just as strong in humans, it is striking to say the least that more and more children under the age of two already have obesity. This would seem to indicate that changes in the body had already taken place early in their development. There are theories based mainly on epidemiological evidence that suggest a changed environment in the uterus or just after birth could largely determine how many fat cells we produce. And as we saw earlier, this is one of the determinants of our weight later in life. It is also known that babies whose mothers smoked during pregnancy have a lower birthweight, but later in life they have a greater risk of obesity.

In any case, it's unlikely that the increase in obesity at this extremely young age can be attributed entirely to poor diet and lack of exercise. Although a genetic abnormality plays a role for some people – as it does for Karen and Jack – these are relatively rare. Moreover, the genes of an entire population don't change that quickly. However, there are studies that show that the effects of the hormone disrupters can be passed on to subsequent generations without any further exposure to them! For example, pregnant mice who had been exposed to a particular hormone disrupter gave birth to offspring who had more fat cells, and these extra fat cells appeared again in the next generation – without being exposed to hormone disrupters!

Hormone disrupters love fat. They are 'lipophilic' and like to nestle into fat cells. Scientists speculate that people who are overweight seem to be even more susceptible to hormone disrupters. When people have more fat cells, there are more fat cells for the hormone disrupters to nestle into, which allows them to contribute even more to the accumulation of fat. This can result in a vicious circle in which both too much fat and too many hormone disrupters can result in all kinds of diseases.

If this has you hyperventilating in your room, wearing an oxygen mask, and you no longer dare to touch a piece of furniture or plastic, let alone eat something that plastic might have leached into, then we've overshot the mark. What we mainly wanted to demonstrate is that the obesity epidemic that has emerged over recent decades is many times more complex than just a direct relationship with an increase in the number of fast-food chains and the number of screens we sit staring at all day long. Once again, we don't know for certain whether these hormone disrupters actually have an impact on the obesity epidemic, or to what extent. But what is important is that, all around the world, more social, political and scientific attention should be devoted to the effects of hormone disrupters on humans. Further research may show that we need stricter legislation and regulations to protect ourselves and also future generations.

Until then, you can, of course, limit your exposure to hormone disrupters by using porcelain cups and glasses, and not eating or drinking out of plastic packaging that has been heated. You would also do well to avoid tinned foods (hormone disrupters leach into the food from the coating on the inside of the tin) and opt for fresh products instead. And make sure you read labels. If you see 'BPA-free', it means that the (food) manufacturer has not used bisphenol A in their products.

ARE YOUR GUT BACTERIA MAKING YOU FAT?
THE ROLE OF OUR MICROBIOME

We carry another hidden contributor to overweight inside us: our gut bacteria, which influence the way in which our body deals with the foods we put in our mouth. These bacteria are also important for our immune system. Our gut is literally filled with these little creatures. In fact, it is estimated that we have ten times as many gut bacteria as all of our body cells combined. These gut bacteria also contain DNA, and the total of all of our gut bacteria along with the genes in their DNA is referred to as our 'microbiome'. Gut bacteria are important to the proper functioning of the various biological processes in our body, such as digestion and metabolism, and as such they keep us at an appropriate weight.

There are many types of gut bacteria, and they all differ slightly in terms of their size, the number of nutrients they absorb from the gut and what they then do with these nutrients. But, to keep things simple, more than 90 per cent of gut bacteria belong to only two so-called 'phyla': Bacteroidetes and Firmicutes. The extent to which you are able to lose weight is thought to be related to the balance between these two kinds of bacteria. Scientific research has shown that people with obesity have fewer Bacteroidetes bacteria in their body, while the Firmicutes bacteria are present in abundance. It has also been shown that the microbiome of people with obesity is 'calibrated' to extract more energy from the food that passes through the gut. Although this is handy when food is scarce, this is not so helpful in times of plenty. In short, what this means is that you can eat less food and still gain weight.

These unfavourable gut bacteria can also lead to obesity in other ways. Some bacteria produce substances that can cause things like mild inflammation, and, in addition, these substances can also promote weight gain. Our microbiome is transmitted to us when we're born, and determines in part how our body deals with the foods we eat. This also explains in part why some people have a 'natural' tendency to become overweight and others don't. Microbiome researchers at Amsterdam University Medical Center found that people who got a wide range of different gut bacteria at birth have a much lower risk of having obesity later in life. It might sound strange, but it's true: the more kinds of bacteria you have in your gut, the better it is for your health.

Luckily, there are things you can do to help to improve the habitat of these tiny gut residents. For example, eating lots of fibre-rich foods (such as vegetables, fruit, wholewheat bread and pastas, brown rice and oats) activates your beneficial gut bacteria. These foods are also called 'prebiotics'. Because these gut bacteria cause our food to pass through the gastrointestinal tract more quickly, our body has less of a chance to absorb all of the nutrients. And this is ultimately beneficial for your fat mass. Interestingly, these good bacteria also make what are known as short-chain fatty acids, of which butyrate is one. Recently, researchers at Leiden University Medical Center discovered that butyrate can be considered to be a substance that protects against excess fat mass accumulation. Studies in mice have shown that butyrate tends to suppress appetite by sending an 'inhibitory signal' from the gut via the nerves to the hypothalamus, which regulates feelings of hunger. Butyrate was also shown to stimulate fat burning via the internal fat-burner brown fat, albeit to a modest extent. In humans, researchers at Amsterdam University Medical Center have recently found initial evidence that butyrate has beneficial effects on glucose metabolism in lean individuals. However, no such beneficial effect has yet been found in people who already had impaired glucose metabolism, such as in prediabetes or diabetes.

On the other hand, 'bad' foods and antibiotics can have a negative impact on our gut bacteria and can cause us to gain weight. This is a real issue in today's world. We consume large quantities of foods high in sugar and fat, and this creates an unfavourable situation for our gut bacteria in which the number of 'good' kinds of bacteria (the voracious eaters) drops and the number of 'bad' kinds of bacteria actually increases, something that may contribute to the trend towards weight gain we see around the world. Something very similar applies to antibiotics. For years, now, small amounts of antibiotics have been added to animal feed. The main reason for doing this is to protect the animals against infectious diseases, but there is also a side effect to this practice that benefits the producer: the animals also get fatter. And it has not escaped scientists' notice that children are being prescribed antibiotics more and more often and at an ever younger age, and that they are also getting heavier. A large Finnish study has shown that, when children take antibiotics, their microbiome becomes 'less diverse' and starts to resemble the microbiome of people with obesity. The researchers also observed that the more often a child had been prescribed a course of antibiotics, the higher their BMI. This is bad news, considering that being overweight early in life results in a higher number of fat cells for the rest of that person's life. More research needs to be conducted to determine whether the current obesity epidemic can actually be attributed in part to antibiotics. But, just like for so many medications, antibiotics should only be prescribed when absolutely necessary, certainly in the case of children.

A remarkable phenomenon is that 'fattening' bacteria are also transmittable. Researchers at the Center for Genome Sciences and Systems Biology at the Washington University School of Medicine in St Louis, Missouri showed that mice that had been raised in a germ-free environment and were then inoculated with gut bacteria from overweight mice got much heavier than mice that had been inoculated with gut bacteria from lean mice. So it would seem that the tendency to have obesity can be transmitted via gut bacteria. Experiments are currently being conducted using stool transplants (yes, you read that right) to inoculate overweight mice with the beneficial gut bacteria of lean mice to improve their metabolism and lower their weight. This is how stool transplants work: a stool sample is collected from a healthy donor and this is mixed with water, among other things. This mixture can be introduced into the gut through a tube inserted via either the mouth or the rectum. Stool transplants have been carried out successfully in humans for some years to treat clostridium difficile, a stubborn gut infection. These stool transplants are being used in more and more ways. In a recent study, researchers at Amsterdam University Medical Center were even able to show that stool transplants from the stool of healthy people can improve the insulin sensitivity of people with 'unhealthy' sugar metabolism, albeit to a modest degree. In future, we will be hearing more and more about the therapeutic potential of our gut. Although this might seem a little distasteful, it's actually very good news.

CAN A VIRUS MAKE YOU FAT?

The American doctor Richard Atkinson, former chair of the American Obesity Association and the North American Association for the Study of Obesity, makes a fascinating case: he says that the current obesity epidemic might be caused in part by a virus. After the explosive increase in obesity in affluent countries in the 1980s, there is now also a huge increase in obesity in poorer countries. In these countries, this increase can't be attributed only to luxury goods such as TVs, computers, microwaves and fast-food chains, soft drinks, large portion sizes and not enough exercise. But . . . a virus?

There is indeed a virus that might be able to cause obesity, namely the cold virus adenovirus 36. At least a third of people with obesity have this virus, while only around 10 per cent of slender people are infected. In the 1990s, researchers in the United States discovered that, if they infected chickens and mice with this virus, these animals would gain a considerable amount of weight. This was also done with monkeys, who are more similar to humans, and all of them gained weight. What was striking was that the animals didn't eat more or become less active, but gained weight as a result of changes to their metabolism and the way in which their body dealt with food. Researchers discovered that the virus's DNA enters the fat cells and causes them to store more fat and sugar from the blood in the form of fat; it also makes it possible for the number of fat cells to increase. Unfortunately, there is currently no treatment for this if you're already infected, but scientists are working diligently to find one. The good news is that people who have obesity and also have this virus are still generally able to lose weight effectively by having a healthy lifestyle – combined, if necessary, with standard anti-obesity medication. However, these people will remain more susceptible to weight gain if they don't manage to maintain a healthy lifestyle. A vaccine is now being developed for this 'fattening' adenovirus 36 to help prevent obesity. Who knows? Perhaps in future it will be possible to use this innovative approach to help to tackle the obesity epidemic.

AN EPIDEMIC OF HIDDEN CONTRIBUTORS TO OVERWEIGHT

If we take everything into account, one thing becomes very clear. We are dealing with a global obesity epidemic that is difficult to halt. For a long time it was thought that lifestyle factors, such as too much of the wrong foods and not enough exercise, were pretty much the only factors that contributed to developing obesity. And far and away the largest number of policies for combating obesity are focused on these factors – unfortunately without success, because the epidemic persists. You could argue that this is logical, given that our diet is still too unhealthy, we don't exercise enough and we are surrounded with supermarkets filled with unhealthy food, which is more likely to work against us than with us in our attempts to improve our diet. But we now know that countless other factors play a role in weight gain (hormones, genes, stress, our mind, lack of sleep, our biological clock, not enough brown fat, etc.). Many of these things have also undergone changes in our current society, and these have all contributed to the fact that, on average, we are carrying more kilos of fat than in years gone by. For an individual, it usually comes down to an accumulation of factors that contribute to weight gain. And we've shown that there is also an entire world of hidden contributors to overweight that are not taken into account nearly enough. Our intestinal flora can either be calibrated to the wrong setting to start with or changed to this setting, many people use medications that have a 'fattening' effect, and – probably even harder to get your head around – it turns out we're also surrounded by hormone disrupters and viruses that potentially also contribute to the obesity epidemic. So now you know all about where things can go wrong. But there is also good news, because there are solutions for dealing with many of these factors. In the next chapter, we will talk more about what you can do in practice to reach a healthy (or healthier) weight and/or to maintain this.

10

How can we tackle overweight effectively?

SLIMMING DOWN THROUGH BARIATRIC SURGERY: PATTY'S STORY

Dutch singer and TV personality Patty Brard grew up in New Guinea and was skinny as a child. She ate very little – so little, in fact, that in primary school she was sent to the school doctor to see if she might have some kind of tropical disease. But that was not the case. Her mother was at her wits' end, and sent Patty to a friend who was a wonderful cook. But Patty would not be tempted. When Patty was eleven, the family moved to the Netherlands. As an adolescent, she reached a weight that was normal for her age. When she was in her twenties, she entered the world of showbiz, where looks – and so also her weight – became increasingly important. Her singing career took off, and she worked long hours.

She remembers this time well: 'During the day, I hardly ate a thing and worked hard. But at night I would "reward" myself. I enjoyed myself in the evenings, going out for dinner, riding around in limousines, and the alcohol flowed freely too. But I was also stressed. I always had to be sociable and entertaining. And the nights were short.' Getting enough sleep was not her priority.

When she turned forty, she began to notice she was gradually gaining weight. Nothing dramatic. It had quietly crept up on her, one pound at a time. She didn't want to admit it at first. How could that skinny girl she had once been possibly be getting fat? Still, she did her best not to gain too much weight. During the day she continued to eat very little. Or was she eating more than she thought? 'I gradually became more and more obsessed with my weight. I was preoccupied with food. It would start in the morning already, and I'd think: I'll pick up a nice piece of meat from the butcher's in a bit, and then pass by the market for some nuts and cheese . . . Cookbooks were almost the only books I read. I know all of Jamie Oliver's books by heart! And then a TV producer said to me, "Those pink Dolce & Gabbana trousers really don't work, Patty." And I realized it had to do with my weight!'

After that, every remark about her weight landed like a bombshell. And that bombshell went off years later when, at the age of sixty-three, her employer made it clear to Patty that they were looking for someone who had a little more 'vim and vigour' than she did. 'I was angry and deeply hurt at the same time. How dare she say something like that to another woman? Later on, though, I was actually grateful to her for being so brutally honest. I had to take a radically different course. And I knew I couldn't go on the way I had been, dieting all day but then going off it at night and automatically rewarding myself with a special dinner. I needed help. And fast.'

Patty didn't feel there was even enough time for the extensive counselling sessions that would be necessary to change her lifestyle. She hurried down to Belgium and had bariatric surgery – a gastric bypass – and lost a huge amount of weight: a total of around 30 kilos within a few months. She watched her body change, and was incredibly happy with what she saw. And she never felt hungry any more – sometimes she would even forget to eat. She thought back in amazement to the time when she was still 'fat', and wondered how she had managed to live with her old body for so long. She felt the energy flowing through her body, and was able to climb the stairs again without panting and puffing. She could wear her pretty dresses again. And high heels! She had her life back.

Still, it wasn't all sunshine and roses. Bariatric surgery is no quick fix. In fact, she had to put at least as much effort into maintaining her weight as she had before the surgery, though she did realize that her previous efforts had backfired: eating nothing during the day and then having large quantities of food in the evening and all those habitual glasses of wine had kept her stressed out and sleep-deprived. If only she'd known about that beforehand . . .

'What I run up against now is that I can't fill up with food. Now I have kind of a built-in alarm that can go off unexpectedly if I eat too much. For example, one night I went into a restaurant and ordered sushi, the amount I was used to ordering in the past. Delicious! That is, until I began to sweat terribly and started to feel unwell. All of my energy just drained away, and I had to sit there quietly until it passed. I was shocked. Later, I learned that this phenomenon is known as "dumping" – if you eat too fast and too much after bariatric surgery, this can produce lots of complaints. My doctors in Belgium had never mentioned this to me. Later on it happened again, at a sidewalk café with a group of companions on the Spanish island of Ibiza. The vegetable shake I was drinking tasted so fantastic that I drank it down very fast. A little while later, I became unwell, like my blood sugar was extremely low, and disappeared to the toilet for an embarrassingly long time. When I got back, I was met by puzzled looks from the people I was with. I thought to myself, 'What must they have been thinking . . .' Now I'm much better at dealing with this. If I'm away from home I always have a container of oatmeal porridge and some crackers with me so I can eat small amounts on a regular basis. And that helps.'

MAINTAINING A HEALTHY
WEIGHT: PREVENTION

The bariatric surgery that Patty had – and every year, many other people worldwide also have this operation – is a radical solution for obesity. But it is a very effective one, certainly for those who have battled against the kilos for years without result, despite all sorts of different diets, powders, pills and intensive lifestyle programmes. We'll return shortly to this drastic solution for obesity, but first we'd like to tell you about treatment methods that have also been shown to work if used properly.

Of course, it's much better to avoid becoming overweight than to cure obesity, and so prevention is by far the most important issue when it comes to tackling obesity. By the time someone's weight has increased dramatically, no matter what the reason for the weight gain, the body is already so dysregulated that finding the way back is very complex. Just think of all of the many hormones that are being produced by the increased amount of body fat and which are disrupting all kinds of bodily processes, including appetite! Even if this person manages to lose weight, many factors that play a role in the hormonal system and metabolism remain disrupted, and the body is programmed, as it were, to easily gain weight again. Thanks to evolution.

If you're lucky, you have a good set of genes and are able to maintain an appropriate weight, even without much physical exercise and in spite of an unhealthy diet. And you might think to yourself that you can keep things that way. But for how long? If we take a realistic look at ourselves as a society, we have to conclude that the vast majority of people do not adhere to the current guidelines for exercise and recommendations for a healthy diet such as those we discussed in Chapter 2 (see the summary in **Box 11**, or go to this WHO website for more comprehensive recommendations: https://www.who.int/nutrition/pub lications/nutrientrequirements/healthy_diet_fact_sheet_394.pdf?ua=1 or https://www.who.int/nutrition/topics/5keys_healthydiet/en/). In the spirit of full disclosure, as the authors of this book on fat, we have to admit that, when it comes to food, neither of us are saints. Although we do our best to have a healthy diet, we too are capable of overindulging in a box of chocolates on occasion, preferably along with a cup of hot chocolate topped with a generous dollop of whipped cream. Actually, seeing as many countries have an obesogenic environment, it's surprising that 'only' 39 per cent of adults worldwide are overweight! Some people do indeed seem to be protected by good genes or other factors that keep them at a healthy weight. At the same time, it's predicted that 86 per cent of people in the US will have overweight or be living with obesity by 2030 if nothing changes. So there's work to be done!

It's undeniable that our world is filled with foods high in sugar and fat. When you're on the go, just try finding something healthy to eat in places like train stations, petrol stations and shopping centres . . . Colourfully wrapped chocolate bars, the tempting aroma of fresh chips, and cream-filled pastries all seem to be calling your name. And your body responds longingly: I want you! This is how we've been programmed by evolution. Just seeing and smelling these treats is enough to get your body and mind ready for action! So how conscious and voluntary is this reaction, then? Who is responsible for turning things around? The individual, or society? Shouldn't governments help the individual make responsible choices through legislation and incentives?

By the way, with a little effort, it really is possible to find something healthy to eat while you're waiting for your train or paying for your fuel at the petrol station, but you'll have to use a magnifying glass to find a green salad among all of the junk high in sugar and fat. In fact, the default setting should be the opposite: the healthy options are the ones that should appeal to us attractively, tempting our body and mind into making a healthy choice. Nowadays, tactics to intentionally influence unconscious behaviour – also known as 'nudging' – are used more and more often. In simple terms, a nudge is a gentle push in the right direction. As we saw earlier, many of our daily decisions are guided by our automatic, intuitive brain. This is exactly what nudges take advantage of: our automatic unconscious behaviour. You would think that this would be a perfect way for governments to guide us towards a healthy diet. In 2008, the American scientists Richard Thaler and Cass Sunstein published their now famous book *Nudge*, in which they translated classic marketing techniques and insights from behavioural economics to the public sector.

Indeed, there are many successful examples of nudges that have led to an increase in sales of healthy products, such as placing these products at eye level on the supermarket shelves, having sliced fruit in the refrigerator case near the cash register (this led to a thirty-fold increase in sales!), vegetable dishes featuring a humorous or attractive 'spoil yourself' recommendation, and smaller portions of less healthy items. The distance to a snack can also be increased, because then people simply eat less of them. By the way, this also applies at work and at home – in other words, don't put the biscuit tin on your desk, but, if you do want to have one around, keep it in an out-of-the-way cupboard. People find it a lot easier to resist a snack if they have to make too much of an effort to get it. How simple (but subtle) marketing can be! But to tackle the obesity epidemic we will also need to take more rigorous measures on the prevention side. For example, the sugar tax – which we mentioned previously – has been successfully introduced in many countries.

WHAT CAN YOU DO YOURSELF TO MAINTAIN OR ACHIEVE A HEALTHY WEIGHT?

In this book, you've already become acquainted with a whole range of factors that have an effect on your body fat and so also your weight. The various factors will play different roles for different people. For example, 'fattening' medications will play the greatest role for one person, and for another person this will be disruption of the biorhythm from working night shifts. What applies to almost everyone, though, is that a healthy diet and enough exercise are the main preconditions for a healthy body and a healthy weight, something that will come as no surprise. In terms of food, don't make it more complicated than necessary. If every day you eat generous servings of preferably fresh fruits and vegetables, wholegrain products, white meat such as chicken, oily fish once a week, dairy products with no added sugar, a handful of unsalted and preferably untoasted nuts, and also drink water, coffee and tea, you're already doing very well (see **Box 11**). And of course this could also include a piece of cake from time to time to celebrate a birthday or to celebrate life. No one ever got fat from one piece of cake! When it comes to food, the main thing is not to be too rigid, but to consciously choose healthy foods. This behaviour will hopefully become increasingly unconscious until this actually becomes your default setting. It's really not necessary to go along with all kinds of fad diets and eat only quinoa, avocado and chickpeas, or to get rid of all carbohydrates or all fats from your diet. Some carbohydrate restriction does seem to be beneficial if a person's glucose metabolism is already impaired, such as in people with diabetes or prediabetes. This is especially the case when a person needs to administer a large amount of additional insulin to properly process the sugars in the body.

Box 11. A healthy diet, in a nutshell

- Eat a number of servings of fruit and vegetables every day
- Eat legumes regularly
- Avoid industrially produced foods (such as ready-made products and processed meats)
- Eat a handful of unsalted nuts every day

- Eat oily fish once a week
- Use wholegrain products
- Have some dairy (with no added sugar) every day
- Drink water/coffee/tea (not beverages with sugar)
- Drink alcohol in moderation

It will come as no surprise that the beverages we drink also contain lots of fattening ingredients. If you send your child to school every day with a sugary drink, this really adds up. In a study done at Amsterdam University Medical Center, 641 children between the ages of four and eleven were followed for a year and a half. Every day, half of the children in the study were given a beverage sweetened with sugar, and the other half were given a beverage that did not contain sugar. After a year, the children who had received a sugar-sweetened beverage every day had gained more than a kilo of extra weight compared with the children who had been given a sugar-free beverage. Let this sink in for a moment: this study was conducted over the course of a single school year! If you assume that a child attends primary school for eight years, that's a difference of more than 8 kilos. And they haven't even got to secondary school yet, where an abundance of things like cakes and chocolate milk are available in the school cafeteria!

And, of course, it's always good to be physically active. You develop more muscle mass and you boost your metabolism. Maybe you don't enjoy exercising, but occasionally find yourself on an exercise bike, sweating profusely as you stare at a display that tells you you've only managed to burn off the calories in a single slice of bread after pedalling away for fifteen or thirty minutes. But, at that moment, please realize that the calories on the screen aren't the most important thing! Exercise, and strength training in particular, builds muscle mass, and it's good to know that not only do you burn calories on the spot when you exercise, but, in part because of your increased muscle mass, your body will burn more calories day and night, 24/7. Even when you're at rest! So you'll burn more calories when you're sleeping, with no additional effort! This is because a constant flow of additional nutrients is being supplied to your muscles, which results in less fat being stored. And this immediately lowers the risk of developing diabetes or other conditions. Your cortisol levels also drop when you exercise regularly, and this is accompanied by beneficial effects. What's more, research has shown that the amount of muscle mass and the quality of the muscles are also excellent predictors of healthy ageing. It keeps you both physically and mentally fit!

In terms of exercise, according to the WHO, the weekly recommendation for adults is 2.5 hours of moderate-intensity aerobic physical activity (e.g. brisk walking or cycling) spread out over the course of the week, or do at least 75 minutes of vigorous aerobic physical activity throughout the week, or an equivalent combination of moderate- and vigorous-intensity activity.

In general, the more often and the more vigorous, the better! Try to spread out your exercise over the course of the week rather than doing nothing all week long and then going all out and exercising intensively at the weekend. If you do the latter, you run a much greater risk of injury, and this will only make matters worse. In addition, at least twice a week you should do muscle- and bone-strengthening exercises. There's no harm in having a little muscle pain after doing strength training. The load placed on your muscles produces tiny micro-tears, and this stimulates the production of more muscle mass. This overcompensation results in muscle growth, and that's a good thing! So, next time you have muscle pain, savour it fully and think of all the good work your body is doing in the meantime. You've done your active part, and now, as you lie there in front of the TV, your body is building muscle mass!

And, as mentioned earlier, to keep from gaining weight, it's also wise to get enough sleep and to avoid too much prolonged stress (if possible). And if you're taking prescription drugs that can potentially contribute to weight gain (see the overview in Chapter 9), make sure to keep a close eye on your weight. And talk to your doctor early on if you notice you're getting heavier. Better to prevent obesity than to reverse it!

NEW SCIENTIFIC INSIGHTS YOU CAN IMPLEMENT YOURSELF

Along with having a healthy diet and exercising more, there are other practical things you can do to reach or stay at a healthy weight. We've already discussed a number of these here, such as boosting your metabolism, and thus your energy expenditure, by fidgeting – making little movements, such as tapping your pen or moving your foot, as you sit otherwise quietly in a chair or at your desk. Sitting still for long periods of time can contribute to obesity, and this can lead to diabetes and cardiovascular disease. And you can't compensate for sitting for long periods just by exercising vigorously in the evening. On top of this, it's not easy to exercise if you're severely overweight, because it puts considerable strain on your joints, and people are often very embarrassed by having to wear exercise gear among the other people at the gym, who are often slender. It has been shown that people who do lots of fidgeting during the day are not at increased risk of cardiovascular disease, even though they sit for long periods. A study has shown that, if you take a smart approach to fidgeting, you can increase your energy consumption by around 20 to 30 per cent. So it would seem that making tiny movements – even something as simple as chewing gum – sets all kinds of different gears in motion inside your body, and that this drastically reduces the harmful effects of sitting still. Although fidgeting is to some extent a personality trait, you could give it a try next time you're stuck in a traffic jam or in a dull meeting that's been going on for hours. If you tighten and relax the muscles in your buttocks for a bit, or tap your foot, you won't be bothering anyone. And who knows? You might be able to subtly boost your energy consumption. It certainly can't hurt.

We actually don't know why sitting still for long periods is so harmful to our health or why it leads to an increase in fat mass. The most obvious thing that springs to mind is that we simply use less energy. However, research conducted with laboratory animals has recently provided another possible explanation. Only when the same results are shown to apply to humans will it offer potential new solutions. It turns out that mice and rats have something that resembles a tiny scale in the bones of their legs, which registers how heavy they are. Up until now, the only mechanism known to monitor our fat mass was the fat hormone leptin, which reports back to the brain and tells it how much fat we have. If we have lots of fat mass, leptin sends a signal to the brain to tell it to decrease appetite and increase metabolism. But these laboratory animals were found to have an additional system for communicating the level of weight to the brain. And this system might explain in part why sitting is so bad for us. When standing, this 'scale in the legs' gives the correct weight. When sitting, this internal scale naturally measures a lower weight and sends this to the brain, and as a result appetite increases and metabolism slows down. This is fascinating new research. Experiments have even been conducted in which mice and rats had to carry a small weight on their back for a short period of time. The result? They lost an amount of weight equal to that of the small weight they were carrying! Further research will have to show whether this mechanism also works this way in humans. If it does, we might well be able to lose weight by fooling the 'weight thermostat' in our brain by carrying around a heavy backpack from time to time!

Along with exercise and fidgeting, there's another smart way to boost your metabolism, which we talked about in Chapter 6: you can expose yourself to low temperatures. This makes your brown fat more active. Spending just two hours a day in a room at 17°C results in a decrease in fat mass while preserving muscle mass. And this is the ideal way to lose weight in a healthy fashion, because it's important to maintain your muscle mass for metabolism. Stimulating your brown fat doesn't have to cost a thing. On the contrary – by lowering your thermostat by one degree, you'll even save money. And if you'd like to exercise, do this outside for the most part. Exercising outdoors for fifteen minutes at 4°C will do you more good than exercising in a heated gym, because then not only will your muscle mass help to boost your metabolism, but so will your brown fat! Eating capsaicin, a substance found in red peppers, affects your brown fat in similar ways. So, if you like spicy food, go ahead and have a few more red peppers!

OVERWEIGHT IS ABOUT MORE THAN JUST WHAT YOU EAT

When it comes to reducing overweight, the first step is to determine exactly why a person is overweight. In most cases, obesity is the result of an accumulation of various factors that contribute to weight gain. In some cases, weight gain can be explained by an underlying disease. It should be noted that organizations including the WHO and the American Medical Association (the largest organization for medical doctors in the United States) define obesity (so, a BMI of 30 or higher) by itself as a disease. And for good reason, when you consider the adverse biological changes that take place in the body when too much fat is present, as we saw in Chapter 4. It's unfortunate that medical professionals often immediately start by giving a treatment recommendation without first looking at what caused the overweight. Because many people, including doctors, believe that weight gain is caused only by what a person puts into their mouth, they try to solve the problem by bringing energy metabolism into balance. So the standard recommendation is usually 'eat less and exercise more'. Problem solved!

This is one of the biggest misconceptions about obesity. Doctors, health professionals, policymakers, politicians, the public at large: we oversimplify the problem. And this is one of the reasons the obesity epidemic has not been halted. To do this, we need to look beyond convenient, one-sided solutions. Now that you have been able to see for yourself all of the different factors that have an effect on our body fat, you also know that the problem of obesity is not a simple one. Although we won't deny that a healthy diet and getting enough exercise are important, with obesity, so much has already been disrupted in the body that it becomes extremely difficult to lose weight just by eating less and exercising more. What has happened is that the body now has a different 'set point', whereby things including all sorts of hormones cause the higher body weight to become the new norm and will make every effort to make sure it remains at that weight.

Having slight overweight is a different story. Not all of these hormonal disruptions are yet involved, so there is a much greater chance of losing weight successfully and also keeping it off, either with or without professional help. So, with obesity, the first step is to look for the causes, for example, by visiting a general practitioner, a specialist in internal medicine, or a paediatrician. But you can also explore for yourself whether there might be factors present that contribute to your weight gain. You can also consider such factors if you're happy with your weight and would like to keep it that way. In this case, categories 1 through 3 are the ones to pay attention to.

THE FOLLOWING SIX CATEGORIES OF FACTORS OR CAUSES CAN CONTRIBUTE TO (SEVERE) OVERWEIGHT:

1 Lifestyle-related factors

- Diet. How is my diet? Do I eat healthy foods like those mentioned by the WHO: '5 keys to a healthy diet' (https://www.who.int/nutrition/topics/5keys_healthydiet/en/), or do I (with some exceptions) routinely choose high-calorie/unhealthy food items (such as having one or more soft drinks every day instead of water)? Or have I been on crash diets that have contributed to weight gain through the yo-yo effect? Do I eat relatively more food in the evening and/or during the night than during the day?

- Exercise. Do I exercise enough, according to the guidelines? And can I also get additional exercise in smaller ways by doing things like taking the stairs more often, fidgeting (tapping my foot or a pen), or chewing (sugar-free) gum? And do I sit for too many hours in a row? At work, could I get up more often (preferably every half hour) to walk to the printer, get coffee, etc.?

- Sleep. Do I sleep for around seven or eight consecutive hours a night? Do I snore loudly and does my breathing stop briefly? Do I ever work night shifts or is my sleep/wake cycle otherwise disrupted?

- Alcohol. Do I drink too much alcohol? You may check whether you are a heavy or even excessive drinker at the CDC website: https://www.cdc.gov/alcohol/fact-sheets/alcohol-use.htm.

- Smoking. Have I just quit smoking? Although people tend to gain weight after they quit smoking, the health benefits of quitting far outweigh this!

- Sociocultural background. Does the culture in which I live keep me from having a healthy lifestyle? For example, because there is always an abundance of (unhealthy) foods, and a lavish array of food is seen as a form of hospitality?

2 Mental factors

- Depressive symptoms. Do I constantly feel so low that I can no longer enjoy things?
- Stress. Do I feel chronically stressed? Am I able to relax enough? Do I have chronic pain (a form of physical stress)?
- Eating binges. Do I have a problem with binge eating (with or without vomiting)? This can indicate bulimia nervosa or binge eating disorder.

3 Medication

- Did my weight go up after I started taking a new medication, or after the dosage increased of a medication I was already taking (and for which weight gain is a side effect)? Medications with this side effect include the following:
- Corticosteroids (tablets, injections or applied locally in large amounts for a prolonged period of time) • Beta blockers (metoprolol, propranolol) • Antidepressants (mirtazapine, citalopram, paroxetine, lithium) • Antipsychotic drugs (olanzapine, risperidone, clozapine, quetiapine) • Antiepileptic drugs (carbamazepine, valproic acid, gabapentin) • Antidiabetic drugs (insulin, glimepiride)
- (Neuropathic) painkillers (pregabalin, amitriptyline)

4 Hormonal factors

Do I have one or more of the following symptoms which, if they occur in a certain combination, can indicate a hormonal cause?

• Constipation • Acne (pimples) • A round face • Dry skin • Excessive hair growth on the face and torso • Bulge of fat on the back of the neck • Slow heart rate • Cold intolerance • Menopause • Irregular menstrual periods • Spontaneous bruising • Purplish-red striae (stretch marks) • Muscle weakness • Brown patches on the neck or in the armpits • Erectile disorders • Weight gained during pregnancy that never entirely disappeared

The following are some examples of hormonal changes or diseases: underactive thyroid, sex hormone deficiency, polycystic ovarian syndrome (PCOS), Cushing's syndrome (too much cortisol), growth hormone deficiency. The last two conditions are rare.

5 Abnormalities in the hypothalamus (control centre in the brain that also regulates appetite and metabolism)

Do I have one or more of the following which, if they occur in combination, may indicate a hypothalamic cause?

• Extreme appetite • Neurological abnormalities • Brain injuries in the past • Radiation of or surgery on the head in the past

The following are examples of hypothalamic causes: injury to the hypothalamus; tumour in the hypothalamus. The causes in this category are rare.

6 Genetic factors

Do I have one or more of the following which, if they occur in combination, may indicate a (rare) genetic cause?

• Extreme appetite • Already severely overweight at a young age • Autism • A striking difference in weight when compared with other family members • Mental or motor developmental delays • No weight loss after bariatric surgery (e.g. gastric bypass or sleeve resection) • Physical abnormalities (such as low-hanging ears, eyes that are set very close together or very far apart, extremely strong prescription lenses, high palate)

The following are examples of rare genetic conditions:

Abnormality in the DNA of a single gene (for example, the MC4 receptor, POMC, leptin); syndromes (for example, Prader-Willi syndrome, Bardet-Biedl syndrome, Alström syndrome, 16p11.2 deletion syndrome).

HOW TO LOSE WEIGHT IF YOU ARE OVERWEIGHT OR ARE LIVING WITH OBESITY

After looking at the factors that contribute to being overweight (on your own or in consultation with a doctor), and optimizing these where possible, it's time to start treatment. If you are overweight (a BMI of 25–30), you can follow the lifestyle recommendations at the beginning of this chapter. Professional help is usually necessary if you have obesity. According to the European Association for the Study of Obesity (EASO) guidelines for adult obesity, the preferred first treatment for obesity consists of lifestyle modifications. An effective treatment is a 'combined lifestyle intervention'. This long-term intensive treatment programme offers guidance in the areas of nutrition, physical activity and behaviour change. In this way, a person can lose around 5 to 10 per cent of their body weight over an extended period of time. In the photo on the opposite page, you see Mark, who participated in a combined lifestyle intervention group at the Obesity Center CGG at the Erasmus University Medical Center in Rotterdam. For a year and a half, he received guidance from a dietitian, a psychologist and a physiotherapist. By participating in this intensive group intervention, he managed to maintain a healthy diet (without dieting!) and to exercise more, and he lost over 28 kilos. Now, more than six years

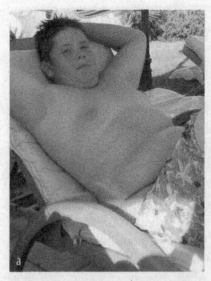

A photo of Mark before (a) and after (b) a combined lifestyle intervention in a group setting. This consisted of recommendations for a healthy (normo-caloric) diet, according to Dutch nutritional guidelines, from a dietitian, exercise coaching by a physiotherapist and cognitive behavioural therapy by a psychologist, for a period of seventy-five weeks in decreasing intensity. (Published with the patient's permission.)

later, he still feels in much better shape. From 2019, certain forms of combined lifestyle interventions will even be covered by the basic health insurance package in the Netherlands. However, this is not the case in most other countries worldwide. If an intensive lifestyle treatment is not producing the desired effect, weight-loss medication or bariatric surgery can be considered.

ANTI-OBESITY MEDICATION

Alas, as yet, there is no such thing as a magical anti-obesity pill (although some dubious internet sites would claim otherwise). There are, however, weight-loss medications that – in combination with lifestyle changes – can result in additional weight loss ranging from around 4.5 kilos to more than 11 kilos. Orlistat and liraglutide are examples of approved anti-obesity drugs. The latter has been in use in many countries for some time now as an antidiabetic medication, and is currently on the market in a higher dose as an anti-obesity medication. However, in many countries it is not yet covered by health insurance. Liraglutide is actually a little sister of GLP-1, the gut hormone from Chapter 5, that curbs appetite and helps with weight loss (especially belly fat). A promising sister drug of liraglutide, which has shown to be even more effective in weight loss, is semaglutide. Another medication that has internationally been approved is a combination pill containing naltrexone and bupropion. These substances are active in the area of the brain that regulates appetite and energy balance. Together, they reduce appetite and the absorption of nutrients, and increase the amount of energy burned. They also reduce our enjoyment of food. In the United States, China and several other countries, in addition to these medications, there are also a number of others on the market that have not yet been approved in Europe, such as lorcaserin and phentermine/topiramate. Interestingly, the effects of all of these medications vary greatly: some people don't lose any additional weight, while others shed many kilos. Although it's not yet known why a medication works for one person and not for another, extensive research is currently being carried out into this. At the moment, if a person taking one of the above anti-obesity medications hasn't lost at least 5 per cent of their weight after three months, it generally doesn't make sense to continue taking that particular medication.

In future, will it be possible to vaccinate our children against obesity in the same way that we now get vaccinations against various kinds of infectious diseases? A futuristic thought, perhaps, but not entirely unrealistic. In Chapter 9 we saw that certain viruses could potentially be hidden contributors to overweight, and vaccinations are already being developed against viruses that are being linked to obesity. What's more, it has been shown that, when laboratory animals (rats, in this case) were vaccinated against the hunger hormone ghrelin, they were less likely to gain weight, produced less fat mass and maintained muscle mass. This is a ground-breaking way to think about obesity prevention.

BARIATRIC SURGERY: A PERMANENT SOLUTION FOR OVERWEIGHT?

Back to Patty's story. She faced numerous obstacles in her battle against the kilos, which went on for decades. In the end, she decided on a radical solution: bariatric surgery. While this is a drastic way to get rid of severe overweight, in combination with lifestyle changes and psychological support it is the most effective treatment for obesity. The numbers don't lie. People lose around 45 kilos on average during the first two years following gastric bypass surgery. Twelve years later, these people have still managed to keep off 35 of these kilos, so it's also effective in the long term. What's more – and this won't come as a surprise when people lose so much fat – some of the complications of obesity completely disappear. They no longer have diabetes (or never develop it), have a reduced risk of cardiovascular disease and cancer, and live longer.

How effective bariatric surgery is depends on the type of operation (see **Figure 8**). The operation Patty had was a 'gastric bypass'. After a gastric bypass, people are only able to eat small amounts of food at a

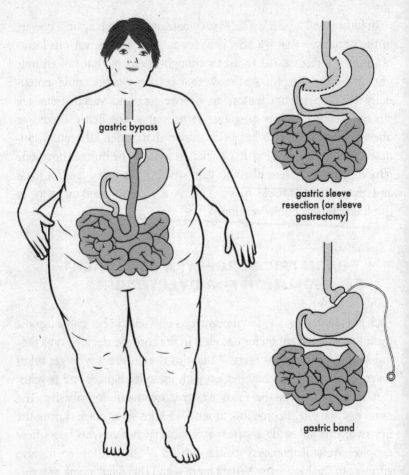

gastric bypass

gastric sleeve resection (or sleeve gastrectomy)

gastric band

Figure 8. Types of bariatric surgery.

time. For someone who really enjoys good food, this is of course a drastic lifestyle change. Although a gastric bypass is major bariatric surgery, it is also one of the procedures that result in the most weight loss. During this surgery, a small part of the stomach the size of an egg is sectioned off and attached directly to the small intestine.

What you then end up with is a tiny stomach. Everything a person eats first enters this tiny stomach, and then goes directly into the small intestine. The first part of the small intestine is also bypassed, which is where the operation gets its name. Then there is the 'gastric sleeve resection' (or sleeve gastrectomy), which involves removing part of the stomach to create a smaller stomach. Here, the connection to the small intestine remains as it was. Finally, a 'gastric band' can also be fitted in people who are extremely overweight. This band makes the stomach entrance physically smaller. This is the least drastic operation, but it is not so effective as the other types of surgery and is therefore performed less often worldwide. In addition, there are also many variations on these operations.

GASTRIC BYPASS BOOSTS
YOUR GUT HORMONES

How does bariatric surgery result in so much weight loss and so many beneficial effects? A great deal of research has been done on this, particularly on the gastric bypass. The most obvious explanation is that people simply can't eat as much. Eating less, especially when it's combined with more exercise, results in a negative energy balance. Fats are released from our body fat and burned by the other body cells. And what happens? Your fat melts before your very eyes. Furthermore, our fat is also less inflamed than before – remember those Pac-Men that make our fat sick? – and this reduces the risk of atherosclerosis and insulin resistance. More than half of the people who had type 2 diabetes at the time of their gastric bypass no longer had it two years after the surgery.

But it gets even more interesting. It turns out that some people no longer have diabetes after just a few days. This is remarkable, because the weight loss only comes later. From this we can conclude that some of the beneficial effects this surgery has on sugar metabolism occur independently of the weight loss. And, indeed, because a gastric bypass results in relatively undigested food going straight into a part of the small intestine that is further along – the first part of the small intestine having been 'bypassed' – many changes take place.

In one of them, the gut hormones mentioned earlier come into play. Somehow, after a gastric bypass, more of the gut hormones GLP-1 and PYY are produced, with all of the beneficial effects this brings with it. People recover quickly from their diabetes because GLP-1 has beneficial effects on sugar metabolism (the pancreas releases more insulin and the organs also became more sensitive to the effects of insulin). But, what's more, after a gastric bypass these gut hormones cause people to feel full faster.

A gastric bypass is major and also drastic surgery. Isn't there a simpler way to boost our gut hormones? Unfortunately, a gastric band – which is the alternative to physically reducing stomach capacity – does not lead to an increase in the beneficial gut hormones GLP-1 and PYY, and people continue to feel much hungrier. And that's while they can't eat as much, because the stomach has become smaller. A highly unpleasant combination.

A more recently developed method for tackling obesity is 'burning off' the gastrointestinal mucosa. This method was first used in humans in 2016. Since the end of the twentieth century, the Brazilian doctor and researcher Manoel P. Galvao Neto has pioneered a wide range of techniques in the gastrointestinal tract to help people with obesity lose weight. He discovered that burning away the gastrointestinal mucosa in the duodenum (the section of gut that comes right after the stomach) mimics the effect of bariatric surgery to a certain degree. The blood sugar levels of people suffering from both obesity and diabetes dropped considerably after only a few days. They also lost several kilos in the period that followed. Here is the idea behind this: there are indications that the gastrointestinal mucosa of people with diabetes is abnormally thick, and that this disrupts sugar metabolism in the body. Techniques that ensure that the mucus lining of the duodenum does not come into contact with the food that enters it result in improved sugar regulation. Burning away the surface of this thin gastrointestinal mucosa changes the contact between the gut wall and nutrients. This could have all kinds of beneficial effects on metabolism, probably due to the effect on gut hormones. We don't yet know the exact mechanism that's involved or the long-term effects, and further research is required.

Right now, research is being carried out on the other ways a gastric bypass has beneficial effects on our metabolism. One of these is that it might alter the composition of bile. Bile is made by the liver, and is made up of water and bile acids, among other things. These bile acids make it easier for fats to be digested in the duodenum. At the end of the gut, the bile acids are reabsorbed and enter the bloodstream. What stands out is that, after a gastric bypass, the amount of bile acids in the blood rises. And this turns out to be very beneficial! In recent years, bile acids have been shown to do much more than just digest fats in the gut. They can bind to special 'bile acid receptors' on many organs, and influence cell metabolism there. And now things get even more interesting. It turns out that these bile acid receptors are also found on our beneficial brown fat. When healthy young women were given bile acid supplements in tablet form for two days, their brown fat became more active and their metabolism speeded up. Perhaps the metabolism of people who have had a gastric bypass also speeds up due to an increase in the amount of bile acids.

A gastric bypass also leads to changes in the composition of our microbiome, the tiny residents of our gut. When the microbiome of mice or humans who had undergone a gastric bypass was transferred to control mice (using a stool transplant), this led to weight loss, loss of fat mass, and beneficial effects on the metabolism of these mice. This would seem to indicate that positive changes take place in the microbiome after a gastric bypass. To what extent these results also apply to humans remains to be seen.

LIFE AFTER BARIATRIC SURGERY

As Patty experienced first-hand, even though bariatric surgery has many beneficial effects on metabolism and makes it possible to lose a large amount of weight, the impact of the surgery should not be underestimated. The road following bariatric surgery is often long and intense, and certainly not always a walk in the park. The fact that people are suddenly only able to eat small amounts of food often hits them hard. This can put them under a great deal of psychological stress, and some people may begin to regret they ever started down this path. Also, the problems that underlie the overeating (such as psychological causes) are not resolved simply by having bariatric surgery. This is why, before a person has bariatric surgery, they undergo a thorough psychological examination and are offered psychological help so that they can get the most out of the surgery in both physical and psychological terms. Before the surgery, all aspects of a person's lifestyle will usually be assessed.

As with any abdominal operation, the short-term risk of bariatric surgery is the risk of the surgery itself (the risk of infection or bleeding). In the long term, people can develop vitamin and mineral deficiencies because a piece of the gut is 'missing', as well as gallstones, narrowing or blockage of the gut or where the stomach opens into the small intestine, and excess skin due to rapid weight loss. And the problem of dumping – which Patty still experiences – also occurs frequently, particularly after gastric bypass surgery. If you eat or drink too much too quickly, you may suddenly become nauseous or experience abdominal pain or diarrhoea within half an hour after eating. This happens when an excessive amount of food suddenly enters the small intestine, and this highly concentrated food pulls a large amount of fluid (litres!) from the bloodstream into the gut. This is known as 'early dumping'. There is also a phenomenon known as 'late dumping'. Around an hour and a half to two hours after eating, you can start to feel dizzy, sweat profusely and even faint, because insulin is still being produced even though the blood sugars have already disappeared from your bloodstream. This is how a person who has diabetes and who takes medication for this feels when their blood sugar is too low.

Finally, if we look at all of the aforementioned obesity treatments – whether surgery, a combined lifestyle intervention, or medication – we can conclude that much more research is needed to be able to understand which treatment is best suited to one individual, and which to another. This is essential in order to ultimately work towards providing personalized obesity treatment. In the end, the reason why someone has obesity differs from person to person. There are usually multiple factors present that contribute to weight gain, and a person with obesity will likely have an inborn predisposition to become overweight in the first place. So we will always first need to look at the underlying causes and contributing factors, optimize these factors where possible, and only then initiate a weight-loss intervention. Although the road to a healthy weight is often very long and hard, as we saw with Patty, it can have a life-changing and sometimes even life-saving effect.

11

Fat-shaming and the psychological consequences of obesity

DIARY OF A 'FATTY': ASHA'S STORY

Asha is a journalist and grew up in East Groningen, a rural region in the northern part of the Netherlands. She has no brothers or sisters. Both of her parents work as psychiatric nurses. And both her mother and her father are overweight. Asha has been chubby since she was a baby, and struggled with her weight throughout her childhood. 'When I was ten, my father went on a diet of slimming shakes. The whole family joined in. After about a year and a half, we switched back to "normal" food, paying close attention to having a healthy diet. I got lots of exercise: swimming, gymnastics, rollerblading and plenty of cycling. I was always hungry.' Asha's mother sent her to school with only a sandwich and an apple, even though she was growing fast, which meant her hunger never went away. Luckily, there was a nice boy at school who always had extra sandwiches, which he would share with her.

When she was ten years old, Asha had a very upsetting experience, the first one that was connected to her being overweight. She had a crush on one of the boys in her class, and he got wind of this in the playground. One of his friends reacted immediately: 'Yuck, Asha has a crush on you? That fatty? I think that's gross!' Asha, who loved to do gymnastics in the playground during breaks, was hanging upside down on the horizontal bar right next to these boys and heard the whole thing. This didn't seem to deter them. On the contrary – it seemed like they wanted to hurt her deliberately.

'At that moment, I learned that my body could trigger disgust, even in someone I had a crush on. And that, as a "fatty", you're sometimes better off keeping certain things to yourself. That apparently I would be better off waiting until someone noticed me, instead of having any expectations of my own when it came to love. This is what I wrote in my diary: "No one is ever going to want me. I'm a pig. Why am I like this? I'm disgusting."'

But Asha is also smart and funny. Lots of people appreciate her for who she is, and she moves through life with ease. Even so, during her primary school years, there were many moments when she would willingly have traded her intelligence and humour – and much more – for a slender waistline. At that time, there was no 'body positivity' movement that called for the acceptance and appreciation of people of all shapes and sizes. And certainly not in the rural part of the country, where she lived. When she thinks back to her childhood, the only person Asha can remember as fat is Roseanne Barr from TV, who was herself the butt of jokes, which she would also direct at herself. 'No one ever told me you could be beautiful even if you were a little on the heavy side.'

When Asha was fourteen, her parents got divorced. This marked the start of a stressful time, during which she barely ate. This was pretty much the only time in her life that her weight was nearly normal. Nearly. Because, in spite of the fact that she ate very little, she continued to struggle with her weight. In the years that followed, she began to eat normally again, and quickly gained back the weight. It's not that she eats lots of sweets – 'I don't even like them. I've always made sure I have a healthy diet, like I learned at home while growing up.' But she does eat quite substantial servings during each 'regular' meal.

Over the course of her adult life, Asha has noticed that people with obesity like herself are viewed differently, and are even discriminated against. For example, when she applied for a different position at work, she was told one morning that unfortunately no more permanent positions were available. However, when a slender man interviewed that afternoon, he was immediately given a permanent position. 'Was it because he was a man? Or because he wasn't overweight like I was?' Other situations were more telling. 'For example, there was the time my daughter and I were sitting on this little ledge, eating ice cream. It was a beautiful spring day. A woman walked up to me and asked what had possessed me to put away a sugar bomb like that, at my weight. "It's going to give you diabetes!" she shouted. My daughter was sitting next to me and got very upset.' She's also had her share of experiences while travelling on public transport. 'One day, I took the train back home. All the seats were taken, but there was one seat with just a backpack on it. I politely asked the owner if I could sit there. "No," said the man. "I don't want to sit next to such a fat slob." Wow. That really got to me.'

Sadly, Asha's experiences aren't unique. Every day in our consultation room we hear the most poignant stories about the prejudice people with obesity face and the nasty personal insults they have to endure. Asha, who studied psychology and is now a renowned science journalist, eventually even wrote a book about the psychology behind food and overweight along with biologist and fellow science journalist Ronald Veldhuizen. Their book is entitled *Eat Me* (*Eet mij*). The mother of Karen – the girl with the genetic form of obesity from Chapter 3 – also remembers the nasty reactions her daughter sometimes got from strangers. 'One day, we were in line at a fast-food restaurant, which was a big deal, because it's something we otherwise never do. I can still feel the icy stares we got from the people around us. They were like a dagger to my heart. Even so, I stayed put, and ordered a salad for Karen. To my amazement, a somewhat older woman came over to us afterwards and apologized. While standing in line, she had wanted to say something to us as parents, and ask us how we could take such an overweight little girl to a fast-food restaurant. When she saw that Karen only got a salad, she was sorry for what she had almost said, and was glad she had held her tongue. That was a sign of character. But still . . .'

DISCRIMINATION BASED ON OBESITY IS THE FINAL FORM OF SOCIALLY ACCEPTED DISCRIMINATION

Everybody knows that we should not discriminate on the basis of ethnicity, skin colour, sexual orientation, age, gender, or mental or physical disability, but somehow it seems that discriminating against people with obesity is socially acceptable. In recent years, there have been heated discussions on topics such as the wearing of headscarves and refugee policies, and during all of these debates considerable attention has been given to whether discrimination is involved. This shows that there is growing awareness that discrimination still exists, and that we need to do something about it. With obesity, this discrimination is much more unconscious and implicit.

In 2001, two American researchers in the field of clinical psychology, Rebecca Puhl and Kelly Brownell, published the first comprehensive overview of decades of research describing bias against and stigmatization of people with overweight and obesity. These distinguished professors showed that obesity carries a stigma in all areas of life: at work, in public and even in health care. This makes it clear how deeply the unfair treatment of people with obesity has penetrated into our society, and how vulnerable these people are. Countless scientific studies have shown that people who are overweight have less disposable income. In the Western world, obesity is known to be more prevalent among people who don't have much money. There are numerous explanations for this: unhealthy food might be cheaper, and thus purchased more often by people with low incomes. Or the socio-economic class someone comes from determines to a great extent what a person eats and how much they exercise. Moreover, people who are in financial difficulty and have debts might experience chronic stress. And, as you now know, an elevated cortisol level can contribute to overweight. But there is also evidence that obesity itself can, in turn, contribute to a lower income. For example, when a person with obesity applies for a job, they are less likely to get it, as Asha suspected when she applied for a permanent position.

Stuart Flint and his colleagues at the Faculty of Health and Well-being at Sheffield Hallam University conducted research on staff recruitment of people with obesity among a group of 180 men and women; of the people in this group, some had made their weight known (by including a photo) and some had not. It quickly became clear to Flint and his team that, when a candidate's weight was not revealed, they were deemed to be more suitable for a job than when it was known that a candidate had obesity. This was true for both men and women. Interestingly, the researchers also observed that people with obesity were considered unsuitable for all forms of manual work (even for light manual work), standing work and even sedentary (seated) work!

There is probably more to this than just the stereotype that a person with obesity is able to do less physically. Other studies have also shown that all sorts of stereotypical images of people with obesity still exist, for example, that they are sloppy and lazy. A final point of interest about Flint's study is that he also observed that gender was one of the determining factors in assessing suitability for a job, with women consistently getting the short end of the stick. If the job opening in question was for a managerial position, a person with obesity was even less likely to get the job; in our society, obesity is automatically associated with being 'less successful'.

THE OBESITY STIGMA IN HEALTH CARE

When you're training to become a medical specialist, you sometimes hear things you'd rather not hear. When we started working in the hospital as medical doctor and researcher, we couldn't believe our ears the first time we heard the letters 'FDD', an abbreviation tossed about by some colleagues. It turned out to be a common abbreviation for 'Fat Dumb Diabetic'. If you know that this is how your doctor sees you, you won't be so comfortable during an office visit.

What this abbreviation makes clear is that those who work in health care also tend to believe that obesity is entirely the person's own fault because they eat too much or don't get enough exercise. Little or no account is taken of genetic predisposition, medication use, stress, lack of sleep, or other factors that can contribute to becoming (or remaining) overweight, and which were discussed in detail in Chapter 9.*

Many doctors don't even bring up the subject of obesity in the consultation room! On the one hand, it can be difficult or painful to address the subject of obesity, and on the other, there is rarely a ready-made solution available and so it's easier to prescribe medication for diabetes, knee pain or depression instead of discussing the underlying causes. And, if the subject does come up, doctors often just recommend going on a low-calorie diet (and we now know that this one-sided approach is not effective), and then go on to prescribe medication that may lead to weight gain – such as certain diabetes medication and medication for high blood pressure, pain or depression – instead of

---

* And even then . . . If the only reason a person is overweight is because they eat too much, this is still no reason to condemn them or deny them medical care. We don't withhold chemotherapy from someone who has lung cancer because they smoked. And when a professional footballer injures their knee during a match, they also receive proper care, even though they would have known that playing football put them at risk. Isn't it their own fault? When is someone actually 'at fault' when it comes to disease?

seeing whether these medications can be tapered off to make it easier to lose weight. For example, if someone is using 100 units of insulin for diabetes, it's virtually impossible to lose weight through a lifestyle intervention, because insulin actually holds on to fat. So, if a person is serious about adopting a healthier lifestyle, they should talk to their doctor to see whether there is a smart way for them to taper off their insulin use. But, as a doctor, you do need to think of this possibility – and, right now, this doesn't happen nearly enough.

In health care, another problem is that diseases are not always properly identified in people with obesity. Various studies have shown that unconscious prejudices also exist among doctors and other health care professionals, and that, as a result, the care provided to people with obesity is generally not as good as the care provided to those with a normal weight.

Asha talks about her own experiences with this: 'Four years ago, I started having some vague symptoms: tiredness, muscle and joint pain, and dry eyes. My general practitioner immediately assumed that my symptoms were being caused by my obesity, and advised me to get a different mattress. My dry eyes were put down to hay fever, even though it was October at that point and hay fever season had ended ages ago. For my stomach complaints, I was referred to a gastroenterologist, who said, "Come back when you're at a healthy weight, and then we can determine whether your symptoms are due to your weight or if it's something else." Of course I knew that many of my symptoms could be explained by my overweight – it's generally known that obesity can cause joint pain and fatigue – but my wrist joints were also painful, and that didn't seem like something that could be caused by my fat bottom.'

Asha was told she should adopt a healthier lifestyle, even though she'd had a healthy lifestyle for years. In fact, she paid more attention to this than many of the people she knew. She went for an hour-long walk every day, went everywhere by bike, did muscle-strengthening exercises every morning, got up from her chair every half hour when she was writing and had a healthy diet. So how was she supposed to 'just lose some weight'? She was also going to see other doctors – a cardiologist because of chest pain, then back to her general practitioner because of recurrent flu-like symptoms, an immunologist because of the autoimmune diseases in her family – but, again and again, her fatigue and joint pain were attributed to her obesity. 'I felt like I wasn't being taken seriously. By then, my symptoms were having a huge impact on my daily life. Even so, I wasn't scheduled to go back to any of those doctors for the time being, but, a year and a half later, things went wrong. One night, I started passing blood in my urine. Everything was red! My general practitioner sent me straight to the internal medicine specialist, who diagnosed me with a kidney condition. It turned out it was an autoimmune disease! "So you were sick, after all!" the specialist said to me, almost in surprise – as if I'd just been pretending I had symptoms the whole time.'

OBESITY AND DEPRESSION HAVE SHARED BIOLOGICAL CAUSES

Asha is very resilient, and she learned from her mother at a young age that she didn't have to be ashamed of her body. Also, despite her early obesity and other setbacks during puberty, like her parents' divorce, she was not prone to actual depression, although she struggled with tremendous sadness for years. However, for many people, obesity and depression go hand in hand. This might seem logical, considering that many people who are overweight have to somehow manage to deal with the prejudices and shame about their body on a daily basis. Indeed, for many people, the stigma surrounding obesity does result in anxiety, out-of-control eating binges and low self-esteem. And the shame they feel about their body is what leads many people to avoid physical activity, with exercise gear that shows all of their curves, and the bulges of fat that get in the way of their attempts to jog. If someone does muster up the courage to give it a try, exercise is accompanied not only by painful knees, but also by the disapproving looks and comments of others. You can see why they might quickly give up.

When you're an obesity doctor, these are exactly the kinds of distressing experiences you hear about every day in the consultation room. What's more, quality of life is considerably affected by functional limitations: 'My belly gets in the way, so I can't even tie my own shoes.' Or, 'I can't run after my toddler if she wants to play with me – or, worse yet, if she runs towards a canal, since she doesn't know how to swim yet.' Although low mood can easily be explained by these daily inconveniences, it's not the whole story. In recent years, we have learned more and more about the biological relationship between obesity and depression, and perhaps in future this will enable us to better help people with obesity to feel better.

The relationship between obesity and depression has been shown to be a two-way street: four large meta-analyses (studies in which the results from earlier studies are combined to reach more accurate conclusions about a particular phenomenon) show the reciprocal relationship between these conditions. Over time, obesity has been shown to be predictive for developing depression, and vice versa: depression increases the risk of developing obesity. This applies not only to adults, but also to children and adolescents. This link between obesity and depression is not unique to the Western world, and has also been reported in other parts of the world. You might think that this association could perhaps be explained by certain factors that increase the risk of both depression and obesity, such as an unhealthy lifestyle, background, greater age, being single, or using antidepressants that contribute to weight gain. But this did not turn out to be the case!

What is fascinating is that there is a great deal of scientific evidence that shows that obesity and depression are linked in a kind of vicious circle of negative adaptations in our body. This is because obesity and depression share biological causes – for example, certain genes result in both conditions, and various hormonal, metabolic and inflammatory mechanisms cause a person to both gain weight and become depressed.

However, this goes beyond a simple one-to-one relationship, because, just as there are different forms of obesity, there are also different forms of depression, namely 'melancholic depression' and 'atypical depression'. A melancholic depression is characterized by such things as the loss of the ability to experience joy, feelings of worthlessness, a flat mood, psychomotor disorders, insomnia, cognitive disorders, loss of appetite and weight loss. In contrast, what is known as atypical depression is characterized by fatigue, excessive sleepiness, and increased appetite and weight gain, among others. With this form of depression, it is precisely this increased appetite that can lead to obesity, along with the biological changes that accompany obesity, such as elevated inflammation levels and hormonal changes like an increase in the fat hormone leptin. Around 40 to 50 per cent of people who are depressed have experienced a decrease in appetite and/or their weight, and a subgroup of around 15 to 25 per cent experienced an increase in appetite or their weight during their depression.

If you hear the stories of Asha, Karen, Mila, Patty and Jack – and many others – they have a good reason to be down. But how are obesity and depression connected, apart from feeling low due to all of the distressing psychological and physical consequences of being overweight? Both obesity and depression are known to have a strong genetic basis. It's remarkable that a large number of the genes associated with obesity mainly exert their influence in the areas of the brain that govern appetite and energy metabolism (the hypothalamus and pituitary gland) and also in the areas that largely determine our emotions and mood (the limbic system). It appears that our weight and energy metabolism are determined by certain areas of the brain that overlap considerably with the parts of the brain that regulate mood. So, if you're unlucky, you were born with genes that put you at greater risk of developing both obesity and depression.

A second biological link between obesity and depression can be found in our stress system: the hypothalamic-pituitary-adrenal axis that ultimately produces the stress hormone cortisol as its end product. In Chapter 8, we described the physical consequences of extended periods of elevated cortisol. When exposure to cortisol is extreme, such as in the rare Cushing's syndrome, around 50 to 80 per cent of individuals suffer from major depression, which usually rapidly disappears again once the source of the overproduction of cortisol is removed from the body. Cortisol was also shown to be elevated in a considerable number of people who did not have this rare condition, but instead had, for example, lifestyle-related obesity. Something else appears to be happening in these cases. It's conceivable that, also here, a high cortisol level is contributing to depressive symptoms, because a surplus of this stress hormone is known to impair areas of the brain, including its emotional centre. Here, as well, cortisol can contribute to a vicious circle, because high cortisol can lead to a craving for snack foods and thus to more belly fat, and more belly fat can in turn lead to elevated cortisol due to all the enzymes, hormones and inflammatory substances produced in the belly fat. So you can see that an excess of both belly fat and cortisol can contribute to low mood. Unfortunately, we can't solve the problem simply by drastically reducing the amount of cortisol, as in people with Cushing's syndrome. Cortisol is an essential hormone and we can't do without it. In fact, without cortisol we would die, because countless metabolic processes would become severely disrupted.

We shouldn't gloss over the inflammatory substances, because they form a third link between obesity and depression. As we saw earlier, fat cells can produce inflammatory substances over time. And we now also know that obesity in the body is actually a state of low-grade inflammation. This is induced not by a bacterial infection (such as with pneumonia), but by the production of inflammatory substances, particularly by belly fat. But why does this make you feel miserable? These inflammatory substances also have a variety of clever ways of reaching the brain – via chemical messengers in the blood, via nerve fibres and via signals transmitted from cell to cell – where they all exert differential effects. Animal studies have even shown that, in obesity, certain areas of the brain exhibit a kind of inflammatory response, including the areas that affect our memory and mood! So, you could say that the belly fat cells cause a mild inflammation in the brain, precisely in those areas that can also cause us to feel depressed.

You might think that you can take an aspirin as an anti-inflammatory and your depression will vanish. And, indeed, a number of scientific studies have shown that anti-inflammatory medications are able to reduce depressive symptoms. Although the results of these studies were promising, they varied from person to person, and it is not yet clear who would benefit from this treatment and who would not. In future, anti-inflammatory medications will hopefully be shown to be effective for those people who are dealing with both obesity and depression.

In Chapters 3 and 5, we described how the fat hormone leptin and other hormones regulate energy metabolism and appetite by, among other things, providing a way for our fat and our brain to communicate. The inflammation we see in obesity disrupts this leptin signal (known as leptin resistance), even though we so badly need leptin to gauge our fat supply and curb our feelings of hunger on time. One of these inflammatory substances is c-reactive protein (CRP), which can interfere with or inhibit the binding of leptin to the leptin receptor. This keeps the satiety signal from being properly transmitted to the main control tower in our satiety and energy metabolism centre: the hypothalamus. This means that we will continue to have increased appetite and also use less energy. Interestingly, leptin itself also appears to be able to affect mood. In animal studies, leptin was found to have an antidepressant effect! A hypothesis that is currently circulating among obesity scientists, but which requires further study, is that leptin resistance could be a major risk factor for developing depression.

An additional factor that forms a link between obesity and depression is another familiar hormone: insulin. This hormone, which helps to regulate sugar and fat metabolism, often becomes disrupted in people with obesity, which can lead to insulin resistance and ultimately even to diabetes (see also Chapter 4). And, yes, insulin also has an effect on the brain, particularly the area where the limbic system – the brain's emotional centre – is located. We have known for some time that depression and diabetes go hand in hand as well. In the past, it was thought that, just as with obesity and depression, this could be explained by a shared genetic background. However, recent studies have shown that this has less to do with the genes. What have been shown to play a role in the combination of depression and diabetes are environmental factors, including the foods high in sugar that we consume in excessive quantities, day in, day out! So it's a good idea to avoid sugary soft drinks and sweets to prevent low mood and to keep us from feeling miserable.

And finally, one more explanatory factor for the relationship between obesity and depression is contained in our gut microbiome. Sometimes we talk about a 'gut feeling'. Did you know that this is backed by scientific fact? This is because our gastrointestinal tract and our brain are busily communicating back and forth. This communication takes place through what is known as the gut – brain axis, which connects the nervous system in the gut with the central nervous system. The gut flora and what we eat play an important role in this gut – brain interaction system. Increasingly, we suspect that this system is involved not only in the development of obesity, but that it also contributes to the development of a number of psychiatric disorders. For example, our gut bacteria have a direct influence on our brain substances, particularly neurotransmitters like serotonin, one of our 'happiness hormones'. A low serotonin level in the brain can result in depressive symptoms. Studies in mice have shown that probiotics (such as Yakult-like dairy drinks) can have an antidepressant effect and reduce anxiety due to changes in serotonin metabolism. Further research is needed to determine whether this also applies to humans.

Some of the changes in our microbiome are also related to inflammatory substances. Among other things, these inflammatory substances appear to be able to increase the permeability of the gut, making it possible for bacteria and other substances to pass through the gut into the blood, and which can then reach the brain. Ordinarily, the brain is protected by a kind of wall, the blood – brain barrier. But the inflammatory substances, such as those that occur with impaired gut flora in the case of obesity, can cause this barrier to 'leak' and allow these substances to enter the areas of the brain that regulate mood. Researchers sometimes also refer to this in terms of 'leaky gut, leaky brain'.

ANTI-OBESITY TREATMENT AND DEPRESSION

In 2011, researchers at the University of Pennsylvania School of Medicine in Philadelphia investigated the effect various kinds of obesity treatments had on depressive symptoms. Do people become less depressed if they lose weight through a lifestyle intervention, anti-obesity medication, or bariatric surgery? The study showed that nearly all weight loss achieved by adopting a healthier lifestyle did indeed lead to a decrease in depressive symptoms. Modifying the entire lifestyle resulted in much greater improvement of the depressive symptoms than just dieting. What also emerged was that exercise interventions had beneficial effects on mood. Interestingly, with these lifestyle interventions, no association was found between changes in weight and changes in depressive symptoms. This would seem to indicate that low mood improved due to factors other than weight loss itself. Cognitive behavioural therapy was frequently used, and it's possible this might have helped in terms of self-acceptance. People learn that their self-worth is not tied to their obesity. Likewise, these therapies can also help to develop an increased sense of self-control, and help in dealing with the stigmatization of obesity in our society. Some of these kinds of life-style interventions are offered in group settings, and the support of peers and therapists can already have beneficial effects on how people feel.

So, what about people who have had bariatric surgery? After bariatric surgery, people generally lose a lot of weight. For many people, this results in improvements in social relationships, more success with the job market and also better quality of life because they have fewer physical limitations. For Patty, her bariatric surgery immediately led to a new life and even a successful television comeback! Diseases related to obesity like diabetes also become less severe or even disappear. Unfortunately, in the initial period following the surgery, for mental quality of life, only modest improvements were observed in terms of psychosocial welfare and depressive symptoms. Studies that followed people over an extended period have shown that depressive symptoms increased again over time.

Interestingly, there is more and more evidence that the risk of problematic alcohol use increases after bariatric surgery. Researchers at the University of Miami in Florida have a number of explanations for this. Physically, our nervous system responds to overeating in about the same way it responds to certain drugs. And so there might be a shift in addictions, whereby, after the surgery, the overeating is exchanged for an alcohol addiction, as it were. Another potential explanation is that sensitivity to alcohol increases after bariatric surgery. A great deal of research is currently being conducted into these phenomena.

The great Roman poet Juvenal (c. 60–c. 140) gave us the concept of 'mens sana in corpore sano': a sound mind in a sound body. As you can see, when it comes to fat management, mind and body are anything but separate systems. The examples of Asha, Patty, Mila, Jack and Karen all show that, as a person with obesity, you need to stand firm to be able to cope with all of the mental blows you have to endure every day. Fat-shaming turns out to be a daily occurrence!

And so it's high time that as a society we show more respect for those who struggle with their weight, and forever do away with the stigma of the Fat Dumb Diabetic. It takes courage for people to talk about obesity, especially for those who have shared their stories with the outside world through this book. It's important to take your own overweight or that of another person seriously. And it's worth your while to have a good look at all of the factors that can contribute to someone becoming or remaining overweight. If you are overweight or are living with obesity, you'll have a better understanding of what is happening, and tackling the underlying causes where possible could help you to reach and maintain a healthier weight. And, above all, when we become aware of all of the factors that together cause the complex condition of obesity, this can lead to greater understanding. Not just an understanding of what maintains your own weight, but also understanding for the overweight of another. It would be a wonderful thing if we stopped passing judgement on each other for being overweight, and instead supported each other, and where possible helped each other to reach and maintain a healthier weight in a good way. There is a long way to go with this weighty problem, but we can get there if we support each other through thick and thin!

ACKNOWLEDGEMENTS

We owe a debt of thanks to a great many people for helping to bring this book into being. To start with, the people who were willing to share their stories with us. These are your own personal stories, which moved and inspired us, and which we feel we all can learn something from. Some of these are your real names, and some are pseudonyms: many thanks to Jack and Karen (and your parents!), and to Asha, Mark, Mila, Natalie, Patty and Rob for your openness!

Because it's important to us that our book contains information that will be useful to many people and also that this information is scientifically substantiated, we asked a number of authorities on science and medicine to check our book for any factual errors.

Patrick Rensen, PhD; Erica van den Akker, MD PhD; Michiel Nijhoff, MD; Yvo Sijpkens, MD PhD; Aart Jan van der Lelij, MD PhD; Adrie Verhoeven, PhD; Max Nieuwdorp, MD PhD; Jan Hoeijmakers, PhD; Jaap Seidell, PhD; Mireille Serlie, MD PhD; Jeroen Molinger; Jan Apers, MD; René Klaassen, MD; and Emma Massey, PhD: many thanks for reading our manuscript and for your valuable feedback.

We would also like to thank a number of proofreaders who checked to see whether the complex material we wanted to present to a wider audience would also be understandable to those without a medical or scientific background. First of all, a huge thank you to our editors Erik de Bruin and Linda Visser. And at least the same degree of thanks to Julie Maturbongs, Marijke Schiffer, Ruud van der Linde, Monique den Hamer, Carla Jongenengel, Claudia Visser, Jos Boon, Ada Willemstein,

Esther Lankhuijzen, Ilke van der Mark and Anita Groenendijk. In addition to these proofreaders, we also would like to thank our (inter)national colleagues in the field of obesity or nutrition for their inspiring discussions which contributed to the views we expressed in this book, in particular Gerda Feunekes, PhD, and Daniëlle Wolvers, PhD, of the Netherlands Nutrition Centre, Arya Sharma, MD, PhD, FRCPC and our colleagues of the European Association for the Study of Obesity (EASO) and The Endocrine Society.

And finally, our heartfelt thanks to our partners, families and friends for the support we received. You enabled us to devote ourselves to the writing of this book during our scant free hours, and, in part thanks to you, the process was a very pleasurable one!

TRANSLATOR'S ACKNOWLEDGEMENT

Grateful thanks to Dr Maarten Groot for generously sharing his insights.

GLOSSARY

Adiponectin One of the many hormones produced by our fat. In mouse studies, it has resulted in things such as increased sensitivity to insulin, lower risk of diabetes and less cardiovascular disease.

Adrenaline A stress hormone produced by the adrenal glands. The adrenal glands rapidly release this hormone when there is acute mental or physical stress.

Agouti-related peptide (AgRP) A neurotransmitter made in the hypothalamus. It causes the brain to generate a feeling of hunger.

Androgens Male hormones such as testosterone.

Biorhythm The day/night rhythm that is linked to the biological clock that is naturally present in your body and even in all of the cells of your organs.

Bisphenol A (BPA) A chemical substance found in many plastic products, and considered to be a 'hormone disrupter'.

Body mass index (BMI) A measure for the ratio between height and weight. It is calculated as weight (kg)/height $(m)^2$. A BMI between 18.5 and 25 is considered to be a healthy weight; a BMI between 25 and 30, overweight; a BMI between 30 and 40, obesity; and a BMI above 40, morbid obesity.

Carbohydrate (complex) Carbohydrates that contain fibre. They first have to be 'snipped' into separate sugars by digestive enzymes before they can be absorbed by the gut, which causes blood sugar to rise slowly.

Carbohydrate (simple) Glucose and fructose. These can easily be absorbed directly into the blood through the gut, and cause blood sugar to rise quickly.

Cell The smallest unit of an organism, such as a plant or animal. It is made up of a cell nucleus, which contains the DNA, and also organelles, the tiny 'machines' that keep the cell running.

Cholecystokinin (CCK) A gut hormone that slows the emptying of food from the stomach. It also promotes satiety by acting on the brain.

Clock genes These form the code for the production of proteins that are important to our body's biological clock. The first clock genes to be discovered formed the code for these proteins: 'period', 'timeless' and 'doubletime'. These three proteins work together and give the proteins in our cells their twenty-four-hour rhythm.

Corticosteroids A collective name for certain hormones that your body either makes itself (such as the stress hormone cortisol) or medications that resemble cortisol, which is made by the adrenal cortex. Examples of this are prednisone and dexamethasone.

Cortisol The stress hormone that is continually being made by the adrenal gland; additional cortisol is made as a reaction to mental or physical stress. It is involved in all kinds of processes in our body, such as the immune system and sugar metabolism. If there is a high level of cortisol in the body, this leads to increased appetite and belly fat!

Cushing's syndrome A constellation of symptoms and signs that point to excessive amounts of the hormone cortisol (produced by the adrenal gland) in the body. This can be the result of taking medication that contains a cortisol-like substance (known as 'corticosteroids'), such as skin ointments, inhaled medication, tablets and injections. It can also result from the body itself producing too much cortisol.

Delta-9-tetrahydrocannabinol (THC) The active ingredient in cannabis. By activating the endocannabinoid receptor in the hypothalamus, it results in increased appetite and the familiar 'munchies'.

Diabetes (In this book, 'diabetes' always refers to type 2 diabetes.) Impaired blood sugar regulation in which blood sugar is elevated. This usually develops because the tissues of the body become less sensitive to insulin (insulin resistance), which causes the sugar to remain in the blood and so the blood sugar level becomes too high. Also, the pancreas often releases less insulin, which means there is not enough insulin for the tissues to absorb the sugars from the blood. Over time, diabetes can lead to such things as cardiovascular disease, kidney diseases, eye problems and problems with the nerves.

DNA The genetic material present in every cell in the body. DNA contains the code for all of the proteins manufactured in the body, such as the proteins that together make up our muscles and eyes.

Dopamine One of the neurotransmitters in the brain responsible for transmitting signals. It is sometimes referred to as one of the brain's 'happiness hormones'. It plays an important role in our reward system, and thus also in addiction.

Dumping A phenomenon that can occur after a gastric bypass. This occurs if you eat or drink too much too quickly, because an excessive amount of food suddenly enters the gut, which causes a large quantity of fluid to be abruptly pulled from the bloodstream into the gut. This can result in a variety of symptoms, such as dizziness, sweating and even fainting.

Endocannabinoids Fat-like substances produced by the body. They are suspended in the blood and able to bind with 'endocannabinoid receptors' and so enter the cells. These receptors are found in the hypothalamus in the brain, and also on various organs such as body fat and muscle. Together with the receptors, the endocannabinoids form the 'endocannabinoid system', a system involved in appetite, our fat and sugar metabolism, our memory and our reward system.

Fatty acid The part of a triglyceride that can be burned to produce energy. There are all kinds of fatty acids: long-chain, short-chain, saturated and unsaturated.

Fidgeting When you fidget or make movements while sitting (such as tapping your foot or pen).

Fructose One of the 'simple carbohydrates', just like glucose. Fructose is also known as 'fruit sugar'.

Gene A location on the DNA that contains the code for a specific protein, such as a receptor.

Ghrelin A hormone produced by the stomach that is responsible for hunger. This is why it's called a 'hunger hormone'.

Glucagon-like peptide 1 (GLP-1) A gut hormone that curbs appetite and prompts the pancreas to release more insulin. In this way, it also lowers the blood sugar level.

Glycogen Stored sugar. It is made up of large clumps of sugar molecules, and is found in two places: in the liver and in the muscles.

Hormone A chemical messenger that is released into the blood by a hormone gland and which produces effects further along in the body. Hormones do this by binding to what are known as hormone receptors on the target organs. An example of a hormone gland is the thyroid gland. But hormones can also be produced and secreted by other organs, such as the heart and our body fat.

Hormone disrupter A chemical substance that can both mimic and block the functioning of our natural hormones. They are also known as 'endocrine-disrupting chemicals'.

Hypophysis (pituitary gland) A gland in the head, about 1 centimetre in size, located right behind the bridge of the nose. This gland contains various hormone-producing cells, which regulate numerous other hormones in the body.

Hypothalamus An area in the brain that includes the satiety centre, fertility centre, temperature centre and 'master clock' (responsible for our biorhythm).

Insulin A hormone produced by the pancreas that enables the glucose in the blood to be absorbed by body cells. It does this by opening gates, as it were, so that the glucose can enter the cell. Insulin also causes you to retain more fat.

Insulin resistance A situation in which body cells become insensitive to the effects of insulin. As a result, the organs absorb less glucose from the blood, and the blood sugar levels rise.

Kisspeptin A hormone produced in the brain that is responsible for the connection between leptin and the fertility centre in the brain.

Leptin The first fat hormone to be discovered. It is made by fat cells, and, by binding to leptin receptors in the hypothalamus, provides a feeling of satiety. Because the amount of leptin produced by the fat cells is in proportion to the amount of fat stored in the fat cells, it is also considered to be the body's 'fat sensor'.

MC4 receptor A receptor in the hypothalamus that, just like the leptin receptor, is involved in bringing about a feeling of satiety.

Mitochondrion Little 'power station' in the cell; great numbers of these are present in practically every cell of our body. They are responsible for metabolism in the cell.

Monogenic obesity This is a relatively rare type of obesity which is often characterized by a severe early-onset obesity with excessive feelings of hunger, decreased satiety and hormonal disorders. Most types of monogenic obesity are due to mutations in genes of the leptin-melanocortin pathway, which plays an important role in the hypothalamic control of food intake and energy metabolism.

Neuropeptide Y (NPY) A neurotransmitter in the brain that generates a feeling of hunger, among other things.

Neurotransmitters Signalling proteins in the brain responsible for transmitting signals.

Nudging A tactic for intentionally influencing unconscious behaviour. A 'nudge' is a gentle push in the right direction.

Obesity Severe overweight whereby a person's BMI is 30 or over.

Obstructive sleep apnoea syndrome (OSAS) A sleep disorder accompanied by snoring and short pauses in breathing. While sleeping, breathing stops multiple times (sometimes as many as fifty times) per hour. This results in a shortage of oxygen in the blood, and a person wakes up exhausted and tends to fall asleep often during the day. It frequently occurs together with overweight or obesity.

Oestrogens Hormones made primarily by the ovaries in women of childbearing age. They can also be produced by body fat when androgens are converted to oestrogen by the protein aromatase.

Overweight A BMI of between 25 and 30.

Pancreas An organ in the belly that produces various hormones, including insulin. It also produces digestive juices that are used to digest the food in our gut.

Peptide YY (PYY) A gut hormone that is responsible for a feeling of satiety.

PET scan A special scan whereby organs are made visible by the absorption of radioactively labelled substances.

Phthalates Chemical substances that make plastic flexible. They are known as 'plasticizers'. They are also considered to be 'hormone disrupters'.

Protein One of our body's energy supplies, along with glycogen and fat. A protein consists of a combination of various 'amino acids', which can in turn be used as fuel. Proteins are also an important component of muscles and many other organs, and they also form receptors.

Receptor A hormone receptor the hormone binds to. A hormone fits into its receptor the way a key fits into a lock. When a hormone binds to its receptor, which are located in or on cells in the body, this leads to various reactions within a specific organ.

Serotonin A neurotransmitter in the brain that induces a 'happy feeling' and makes you emotionally stable, among other things. Via the MC4 receptor, it also induces a feeling of satiety.

Starch Long chains of glucose.

Thermogenesis Energy expenditure. Total daily energy expenditure is divided into different components, including resting metabolic rate, the thermic effect of food, and activity thermogenesis.

Triglyceride The stored form of fat. It is made up of three fatty acids bound to glycerol.

SOURCES

1 A LOOK AT THE HISTORY OF FAT

The information about Venus of Willendorf came from the article by Antl-Weiser, W., 'The anthropomorphic figurines from Willendorf'. *Wissenschaftliche Mitteilungen Niederösterreichisches Landesmuseum* (2008), 19:19-30.

A lucid overview of the history of how overweight is perceived is contained in Eknoyan, G., 'A history of obesity, or how what was good became ugly and then bad'. *Advances in chronic kidney disease* (2006), 13:421-427.

The discoveries concerning the causes of overweight are superbly set out in Bray, G., 'Obesity: Historical development of scientific and cultural ideas'. *International Journal of Obesity* (1990), 14: 09-926.9.

We obtained information on the second agricultural revolution from Fogel, R.W., *The Escape from Hunger and Premature Death, 1700–2100*. Cambridge, UK, Cambridge University Press, 2004.

A detailed overview of the discovery of the fat cell and all of the important discoveries concerning fat are found in Lafontan, M., 'Historical perspectives in fat cell biology: the fat cell as a model for the investigation of hormonal and metabolic pathways'. *American Journal of Physiology – Cell Physiology* (2011), 302:C327-C359.

For the item on the Fat Men's Club, we visited the website of the New England Historical Society.

2 FAT AS AN ESSENTIAL ORGAN FOR STORAGE

A nice description of fat, sugar and protein metabolism comes from Chapter 46 of the book by Levy, M. et al., *Physiology Fourth Edition*. Mosby, Philadelphia, 2006.

How the fat cell works and the foetal development of body fat can be found in this good review article by Symonds, M.E. et al., 'Adipose tissue and fetal programming'. *Diabetologia* (2012), 55: 1597-1606.

A complete overview of lipodystrophy is contained in the following article: Jazet, I.M. et al., 'Therapy resistant diabetes mellitus and lipodystrophy: leptin therapy leads to improvement'. *Nederlands Tijdschrift voor Geneeskunde [Dutch Journal of Medicine (NTvG)]* (2013), 157(4):A5482.

An overview of the various kinds of sugars and fats can be found on the website of the Netherlands Nutrition Centre (Voedingscentrum).

A study that describes the various effects of fructose and glucose on the liver can be found in Jensen, T. et al., 'Fructose and sugar: a major mediator of non-alcoholic fatty liver disease'. *Journal of Hepatology* (2018), 68: 1063-1075.

Would you like to read more about the difference between the effect of a diet low in fat and a diet low in sugar? See Gardner, C.D. et al., 'Effect of low-fat vs low-carbohydrate diet on 12-month weight loss in overweight adults and the association with genotype pattern or insulin secretion: the DIETFITS randomized clinical trial'. *JAMA* (2018), 319: 667-679 and Savas, M. and van Rossum, E.F.C., 'Beter een vetarm of koolhydraatarm dieet: is dat te voorspellen?' ['Is a low-fat or low-carbohydrate diet preferable and can this be predicted?']. *Nederlands Tijdschrift voor Geneeskunde [Dutch Journal of Medicine (NTvG)]* (2017), 161:D2310.

The study that shows that eating carbohydrates last is beneficial is Shukla, A.P. et al., 'Carbohydrate-last meal pattern lowers postprandial glucose and insulin excursions in type 2 diabetes'. *BMJ Open Diabetes Research & Care* (2017), 5:e000440.

A review on the latest insights on which proportion of foods your diet can best consist of is contained in Ludwig, D.S. et al., 'Dietary fat: from foe to friend?'. *Science* (2018), 362:764-770.

International recommendations for healthy nutrition ('5 keys to a healthy diet') can be found on the following WHO website : https://www.who.int/nutrition/topics/5keys_healthydiet/en/

In some more detail, nutritional information can be found at : https://www.who.int/nutrition/publications/nutrientrequirements/healthy_diet_fact_sheet_394.pdf?ua=1

A nice review article that the Dutch dietary guidelines of the Netherlands Nutrition Centre (Voedingscentrum) are based on can be found in Kromhout, D. et al., 'The 2015 Dutch food-based dietary guidelines'. *European Journal of Clinical Nutrition* (2016). 70:869-878. The instructions Mark (in Chapter 10) received during his combined lifestyle intervention were also based on these guidelines.

The article stating that eating fibre regularly leads to less cardiovascular disease can be found in Reynolds, A., 'Carbohydrate quality and human health: a series of systematic reviews and meta-analyses'. *The Lancet* (2019), 393:34-445.4.

3 FAT AS A HORMONE FACTORY

The information about the discovery of leptin comes from Chapter 1 ('The obese (ob/ob) mouse and the discovery of leptin') of the book *Leptin* by V. Daniel Castracane and Michael C. Henson.

The discovery of leptin deficiency in two children is described in Montague, C.T. et al., 'Congenital leptin deficiency is associated with severe early-onset obesity in humans'. *Nature* (1997), 26: 903-908.

A good overview of the successful treatment of leptin deficiency with leptin can be found in Paz-Filho, G. et al., 'Leptin treatment: facts and expectations'. *Metabolism* (2015), 1:146-156.

The first time a child with leptin deficiency was treated with leptin is described in Farooqi, I.S. et al., 'Effects of recombinant leptin therapy in a child with congenital leptin deficiency'. *New England Journal of Medicine* (1999), 341:879-884.

The role of leptin in preventing the yo-yo effect after weight loss is described in Olivia M. Farr et al., 'Leptin applications in 2015: What have we learned about leptin and obesity?' *Current Opinion in Endocrinology, Diabetes and Obesity* (2015), 22:353-359.

A nice review article on leptin's potential in treating obesity and diabetes can be found in DePaoli, A.M., *Journal of Endocrinology* (2014), 223:T71-T81.

A good recent article with an extensive explanation of the properties of adiponectin is contained in Nigro, E. et al., 'New insight into adiponectin role in obesity and obesity-related diseases'. *BioMed Research International* (2014), 658913.

We obtained information on body composition and menstruation among top-level gymnasts from the following articles: Claessens, A.L. et al., 'Growth and menarcheal status of elite female gymnasts. *Med Sci Sports Exerc.* (1992); 24:755-763; Theintz, G.E. et al.; 'Evidence for a reduction of growth potential in adolescent female gymnasts'. *Journal of Pediatrics* (1993), 122:306-312; and Beunen, G., 'Physical growth and maturation of female gymnasts: influence of selection bias on leg length and the duration of training on trunk length'. *Journal of Pediatrics* (1999), 136:149-155.

Two nice review articles on leptin's role in fertility are Chehab, F.F. et al., 'Leptin and reproduction: Past milestones, present undertakings and future endeavors'. *Journal of Endocrinology* (2014), 223:T37-T48 and Manfredi-Lozano, M. et al., 'Connecting metabolism and gonadal function: novel central neuropeptide pathways involved in the metabolic control of puberty and fertility'. *Frontiers in Neuroendocrinology* (2018), 48:37-49.

Would you like to learn more about the romantic Kisspeptin protein? See Skorupskaite, K. et al., 'The kisspeptin-GNRH pathway in human reproductive health and disease'. *Human Reproduction Update* (2014), 20:485-500.

4 FAT CAN RESULT IN DISEASE AND DISEASE CAN MAKE YOU FAT

Two good review articles on the life cycle of fat: Hyvönen, M.T. et al., 'Maintenance of white adipose tissue in man'. *The International Journal of Biochemistry & Cell Biology* (2014), 56:123-132 and Arner, P. et al., 'Fat cell turnover in humans'. *Biochemical and Biophysical Research Communications* (2010), 396:101-104.

The study that describes the effect of weight gain on fat cell size is contained in Salans, L.B., et al., 'Experimental obesity in man: cellular character of the adipose tissue'. *Journal of Clinical Investigation* (1971), 50:1005-1011.

The most important study to show that the number of fat cells remains constant from around age twenty was published in 2008 by Spalding, K.L. et al. in *Nature*. The title is 'Dynamics of fat cell turnover in humans'.

A review on the effects of sex hormones on the various fat depots can be found in White, U.A. et al., 'Sex dimorphism and depot differences in adipose tissue function'. *Biochimica et Biophysica Acta* (2014), 1842: 377-392.

An extensive review on the differences between belly fat and subcutaneous fat and the risk of metabolic diseases is contained in Schoettl, T. et al., 'Heterogeneity of adipose tissue in development and metabolic function'. *Journal of Experimental Biology* (2018), 221:jeb162958.

The effect of obesity on fertility in women and the effect of weight loss on improving fertility is described in the following two articles: Silvestris, E. et al., 'Obesity as disruptor of the female fertility'. *Reproductive Biology and Endocrinology* (2018), 16:22 and Best, D. et al., 'How effective are weight-loss interventions for improving fertility in women and men who are overweight or obese? A systematic review and meta-analysis of the evidence'. *Human Reproduction Update* (2017), 23:681-705.

The effect of obesity on fertility in men is contained in the following review: Liu, Y. et al., 'Obesity, a serious etiologic factor for male subfertility in modern society'. *Reproduction* (2017), 154:R123-R131.

The study in which Rose Frisch showed that former athletes have a lower risk of developing breast cancer and cancer of the reproductive organs is described here: Frisch, R.E., 'Former athletes have a lower lifetime occurrence of breast cancer and cancers of the reproductive system'. *Advances in Experimental Medicine and Biology* (1992), 322:29-39.

The association between overweight and obesity and developing cancer is described in these two reviews: Allot, E.H. et al., 'Obesity and cancer: mechanistic insights from transdisciplinary studies'. *Endocrine-Related Cancer* (2015), 22:R365-R386 and Berger, N.A. et al., 'Obesity and cancer pathogenesis'. *Annals of the New York Academy of Sciences* (2014), 1311:57-76.

5 HOW DO FEELINGS OF HUNGER AND SATIETY WORK?

An impressive study that shows that we make more than 220 unconscious food choices every day and how our environment influences these decisions was published in 2007 by Brian Wansink and Jeffery Sobal in *Environment and Behavior*. The title of the article is 'Mindless Eating: The 200 Daily Food Decisions We Overlook'.

In their Dutch-language article 'Obesitas: gendiagnostiek of geen diagnostiek?' ['Obesity: genetic diagnostics or no diagnostics?'], paediatric endocrinologists Erica van den Akker and Edgar van Mil describe how diagnostics into 'monogenic' forms of obesity such as those that affect Karen (leptin receptor abnormality) and Jack (MC4 receptor abnormality) can be used in children with obesity, in *Praktische Pediatrie* ['Practical Paediatrics'] (2009).

A moving story (in Dutch) written by a woman who is severely overweight herself and who was diagnosed with a genetic abnormality similar to Jack's is *Knap voor een dik meisje. Het gewicht van gewicht* ['Not bad for a fat girl. The weight of weight.'] (2019). Tatjana Almuli (former participant in the TV programme *Obese*). Amsterdam: Nijgh & Van Ditmar.

Our scientific article about how often genetic obesity can be found in specific groups of children and adults with obesity in the Netherlands was published in 2018 in the *Journal of Medical Genetics* by L. Kleinendorst, M.P.G. Massink, M.L. Cooiman, M. O.H. van der Baan-Slootweg, R.J. Roelants, I.C.M. Janssen, H. Meijers-Heijboer, N.V.A.M. Knoers, H.K. Ploos van Amstel, E.F.C. van Rossum, E.L.T. van den Akker, G. van Haaften, B. van der Zwaag, M.M. van Haelst. The title is 'Genetic obesity: next-generation sequencing results of 1230 patients with obesity'.

A nice review article on ghrelin, the hunger hormone from the stomach, was published in 2007 by R.M. Kiewiet, M.O. van Aken, L. Schepp, Y.P.M. van der Hulst, and A.J. van der Lelij in the *Nederlands Tijdschrift voor Klinische Chemie en Laboratoriumgeneeskunde* [*Dutch Journal for Clinical Chemistry and Laboratory Medicine*], entitled 'Ghreline: van eerste natuurlijke groeihormoon secretagoog tot multifunctioneel peptide' ['Ghrelin: from first natural growth hormone secretagogue to multifunctional peptide'].

A fascinating study that describes the effect of our thoughts on how quickly we feel full and on our hunger hormone ghrelin is contained in the article by Alia J. Crum et al., 'Mind over milkshakes: mindsets, not just nutrients, determine ghrelin response', in *Health Psychology* (2011).

Additional information about the 'sister hormone' of hunger hormone ghrelin, which was unexpectedly found to have a variety of beneficial effects on metabolism, is contained in 'Des-Acyl Ghrelin: A Metabolically Active Peptide' by Delhanty, P.J., Neggers, S.J., van der Lely, A.J. in the book *The Ghrelin System, Endocrine Development Karger* (2013) (Benso, A., Casanueva, F.F., Ghigo, E. and Granata, A., editors).

Internal medicine specialist and scientist Werner Creutzfeldt describes the history of the discovery of our gut hormones, which are able to communicate with our brain, in 'The (pre)history of the incretin effect' in *Regulatory Peptides* (2005).

In 'Cannabinoid pharmacology: the first 66 years', researcher R.G. Pertwee from the University of Aberdeen takes you on a journey through the history of cannabis and the body's own endocannabinoid system, which has been shown to have a variety of effects on our brain and body. In *British Journal of Pharmacology* (2006).

An interesting article on how our brain's reward system works, how this system enables us to overeat and how we can exercise more control over it is 'Obesity and the Neurocognitive Basis of Food Reward and the Control of Intake', in *Advances in Nutrition* by Hisham Ziauddeen et al. (2015).

Additional scientific information about why some people have an extremely strong need for carbohydrates and how this relates to emotional disorders and obesity can be found in 'Neurobiologic basis of craving for carbohydrates', by Ventura, T. et al., in *Nutrition* (2014).

The effects of 'happiness hormone' serotonin, which plays an important role in addiction and can curb appetite, is described in the article 'The role of serotonin in drug use and addiction' by Müller, C.P. and J.R. Homberg, in *Behavioural Brain Research* (2015).

This scientific article describes a medication that is on the market in the United States that curbs appetite and has an anti-obesity effect through the reward system: 'Role of impulsivity and reward in the anti-obesity actions of 5-HT2C receptor agonists', in *Journal of Psychopharmacology* (2017), by G.A. Higgins, F.D. Zeeb and P.J. Fletcher.

6 MARVELLOUS METABOLISM

A review on the properties of brown fat and various medications and food components that have an effect on this can be found in Ruiz, J.R. et al., 'Role of human brown fat in obesity, metabolism and cardiovascular disease: strategies to turn up the heat'. *Progress in Cardiovascular Diseases* (2018).

The case study of Barbara was based on the following case study about a patient with a hibernoma: Gadea et al., 'Hibernoma: a clinical model for exploring the role of brown adipose tissue in the regulation of body weight?' *The Journal of Clinical Endocrinology & Metabolism* (2014), 1:1-6.

A review on 'non-exercise activity thermogenesis (NEAT)' is described here: Levine, J.A. et al., 'Non-exercise activity thermogenesis (NEAT): environment and biology'. *American Journal of Physiology – Endocrinology and Metabolism* (2004), 286:E675-E685.

The study on the effects of replacing sitting with more standing and walking is described in Duvivier, B. et al., 'Minimal intensity physical activity (standing and walking) of longer duration improves insulin action and plasma lipids more than shorter periods of moderate vigorous exercise (cycling) in sedentary subjects'. *PLoS One* (2013).

The effects of foods on brown fat are described in Yoneshiro, T. et al., 'Tea catechin and caffeine activate brown adipose tissue and increase cold-induced thermogenic capacity in humans'. *American Journal of Clinical Nutrition* (2017), 105:873-881.

A nice Japanese study that showed that six weeks of exposure to mild cold leads to loss of fat mass and also that eating capsaicin (found in red peppers) every day activates brown fat in healthy young men was published in 2013 by Yoneshiro et al., in the *Journal of Clinical Investigation*.

7 FAT AND OUR BIORHYTHM

More information on the ground-breaking research into the biological clock for which three American researchers (Jeffrey C. Hall, Michael Rosbash and Michael W. Young) won the 2017 Nobel Prize in Physiology or Medicine is contained in this video by National Public Radio of the prize's announcement: https://www.npr.org/sections/thetwo-way/2017/10/02/554993385/nobel-prize-in-medicine-is-awarded-to-3-americans-for-work-on-circadian-rhythm.

A nice review on how lack of sleep can result in increased appetite, changed metabolism and weight gain, and the effects of various sleep interventions, were described in 2014 by Arlet Nedeltcheva and Frank Scheer in 'Metabolic effects of sleep disruption, links to obesity and diabetes', in *Current Opinion in Endocrinology & Diabetes and Obesity*.

The interesting study in which researchers extended the sleep duration of people who were usually short sleepers and then investigated whether this had beneficial effects on their diet was published by Haya, K. Al Khatib et al., from King's College London, in 2018, in *The American Journal of Clinical Nutrition*: 'Sleep extension is a feasible lifestyle intervention in free-living adults who are habitually short sleepers: a potential strategy for decreasing intake of free sugars? A randomized controlled pilot study'.

This is the large British study that showed that women who sleep in a bedroom that is not completely dark are heavier than women who sleep in a dark bedroom: 'The Relationship Between Obesity and Exposure to Light at Night: Cross-Sectional Analyses of Over 100,000

Women in the Breakthrough Generations Study', by Emily McFadden et al., in *American Journal of Epidemiology* (2014).

This article by Muscogiuri, G. et al., provides a clear overview of the reciprocal associations between sleep and obesity – how lack of sleep can increase our body weight, but also how obesity and unhealthy diet can result in poorer sleep quality: 'Obesity and sleep disturbance: the chicken or the egg?', in *Critical Reviews in Food Science and Nutrition* (2018). Associations between sleep and obesity are also described in the article 'Sleep Duration and Obesity in Adults: What Are the Connections?' by Theorell-Haglöw, J. and Lindberg, E., in *Current Obesity Reports* (2016).

Research that showed an increased risk of developing cardiovascular disease if you have both obesity and OSAS was described by Jean-Louis Girardin et al. in 2008, in the article 'Obstructive Sleep Apnea and Cardiovascular Disease: Role of the Metabolic Syndrome and Its Components, in *Journal of Clinical Sleep Medicine*.

This article by C.L. Grant et al. describes how important timing is in terms of when you eat certain foods and that eating during the night can disrupt your blood sugar regulation: 'Timing of food intake during simulated night shift impacts glucose metabolism: A controlled study', in *Chronobiology International* (2017).

This is an important review article on the effects on health of specific eating patterns, such as skipping breakfast, intermittent fasting and the number and timing of meals: 'Meal Timing and Frequency: Implications for Cardiovascular Disease Prevention A Scientific Statement From the American Heart Association', by Marie-Pierre St-Onge, in *Circulation* (2017).

In 2011, Priya Sumithran et al. published a ground-breaking study in the prestigious New England Journal of Medicine on the short- and long-term effects of a short very-low-calorie diet on our hunger and satiety hormones entitled 'Long-Term Persistence of Hormonal Adaptations to Weight Loss'. This clearly showed how part of the yo-yo effect (gaining weight back after a diet) can be explained.

If you would like to read more about the longer-term outcomes of the well-known American competition *The Biggest Loser*, you can read the article 'Persistent metabolic adaptation 6 years after "The Biggest Loser" competition', by E. Fothergill et al., published in 2016 in *Obesity*.

In this article, you can read more about the diet trend 'intermittent fasting', whereby periods of fasting are alternated with periods of eating: 'Intermittent Fasting: Is the Wait Worth the Weight?' by M.C. Stockman et al., in *Current Obesity Reports* (2018).

You can find more information about another diet trend, 'time-restricted fasting' (whereby the number of calories a person is allowed to eat is not decreased, but the time period in which food is eaten is drastically reduced), in the article 'Time-restricted feeding for prevention and treatment of cardiometabolic disorders', by G.C. Melkani and S. Panda, in *Journal of Physiology* (2017). The American researcher Panda also wrote an interesting article with S. Gill, which was published in 2015 in the leading journal *Cell Metabolism*. In the article 'A Smartphone App Reveals Erratic Diurnal Eating Patterns in Humans that Can Be Modulated for Health Benefits', about the eating patterns of healthy adults, they also made associations with health effects.

8 HOW DOES STRESS CAUSE OVERWEIGHT?

The following articles describe the science behind the association between stress and obesity: 'Obesity and cortisol: New perspectives on an old theme', by E.F.C. (Liesbeth) van Rossum, in *Obesity* (2017) and 'Stress and Obesity', by A.J. Tomiyama, in *Annual Reviews* (2018).

In this article, we describe why some people are more susceptible than others to the negative effects of chronic stress on body weight: 'Stress and Obesity: Are There More Susceptible Individuals?' by Van der Valk, E.S., Savas, M., and Van Rossum, E.F.C., in *Current Obesity Reports* (2018).

You can find more information on Cushing's syndrome, which was what Mila was found to have, at https://www.hormone.org/diseases-and-conditions/cushing-syndrome.

This article describes how we studied the relationship between obesity and the stress hormone cortisol in adults using a relatively new method to measure long-term cortisol levels in scalp hair: 'Long-term cortisol levels measured in scalp hair of obese patients', by Wester, V.L., Staufenbiel, S.M., Veldhorst, M.A., Visser, J.A., Manenschijn, L., Koper, J.W., Klessens-Godfroy, F.J., Van den Akker, E.L.T., and Van Rossum, E.F.C., in *Obesity* (2014).

Our research on the association between cortisol and obesity in children is described in 'Long-term glucocorticoid concentrations as a risk factor for childhood obesity and adverse body-fat distribution', by Noppe, G., Van den Akker, E.L.T., de Rijke, Y.B., Koper, J.W., Jaddoe, V.W., and Van Rossum, E.F.C., in *International Journal of Obesity* (2016).

We described the relationship between chronically elevated cortisol and increased risk of cardiovascular disease in 'High long-term cortisol levels, measured in scalp hair, are associated with a history of cardiovascular disease', by Manenschijn, L., Schaap, L., Van Schoor, N.M., Van der Pas, S., Peeters, G.M., Lips, P., Koper, J.W., and Van Rossum, E.F.C., in *Journal of Clinical Endocrinology and Metabolism* (2013).

The article 'Use of hair cortisol analysis to detect hypercortisolism during active drinking phases in alcohol-dependent individuals', by Stalder, T., Kirschbaum, C., Heinze, K., Steudte, S., Foley, P., Tietze, A. and Dettenborn, L., in *Biological Psychology* (2010), describes the association between excessive alcohol consumption and cortisol.

More information on harmful alcohol consumption worldwide is made available online by the WHO (https://www.who.int/health-topics/alcohol#tab=tab_1).

9 HIDDEN CONTRIBUTORS TO OVERWEIGHT

The association between medication that contains a type of stress hormone (namely, corticosteroids) and body weight is described in the following scientific articles: 'Systematic Evaluation of Corticosteroid Use in Obese and Non-obese Individuals: A Multi-cohort Study' by Savas, M., Wester, V.L., Staufenbiel, S.M., Koper, J.W., Van den Akker, E.L.T., Visser, J.A., Van der Lely, A.J., Penninx, B.W.J.H. and Van Rossum, E.F.C., in *International Journal of Medical Sciences* (2017), and 'Associations Between Systemic and Local Corticosteroid Use With Metabolic Syndrome and Body Mass Index' by Savas, M., Muka, T., Wester, V.L., Van den Akker, E.L.T., Visser, J.A., Braunstahl, G.J., Slagter, S.N., Wolffenbuttel, B.H.R., Franco, O.H. and Van Rossum, E.F.C., in *Journal of Clinical Endocrinology and Metabolism* (2017).

In addition, in 2015, an article was published that clearly shows that also locally applied medications containing corticosteroids can produce effects throughout the body, which can also suppress the body's own adrenal function: 'Adrenal Insufficiency in Corticosteroids Use: Systematic Review and Meta-Analysis' by Broersen, L.H., Pereira, A.M., Jørgensen, J.O. and Dekkers, O.M., in *Journal of Clinical Endocrinology and Metabolism*. More information on the relationship between obesity and asthma is contained in the following articles: 'The influence of obesity on inflammation and clinical symptoms in asthma' by Gruchała-Niedoszytko, M., Małgorzewicz, S., Niedoszytko, M., Gnacińska, M. and Jassem, E., in *Advances in Medical Sciences* (2013), and 'Underdiagnosis and overdiagnosis of asthma in the morbidly obese' by Van Huisstede, A., Castro Cabezas, M., Van de Geijn, G.J., Mannaerts, G.H., Njo, T.L., Taube, C., Hiemstra, P.S. and Braunstahl, G.J., in *Respiratory Medicine* (2013).

In *Current Opinion in Pulmonary Medicine* (2016), C.S. Ulrik has written about the beneficial effects of weight loss on asthma in people who have both obesity and asthma in the article 'Asthma and obesity: is weight reduction the key to achieve asthma control?'

An overview of the medications that can have even more weight gain as a side effect, and to what degree, is contained in our article 'A comprehensive diagnostic approach to detect underlying causes of obesity in adults', by Eline S. van der Valk, Erica L.T. van den Akker, Mesut Savas, Lotte Kleinendorst, Jenny A. Visser, Mieke M. van Haelst, Arya M. Sharma and Elisabeth F.C. (Liesbeth) van Rossum, in *Obesity Reviews* (2019).

This article by M. Dayabandara et al. describes the mechanisms of how certain antipsychotics can lead to weight gain and also ways to counteract this effect: 'Antipsychotic-associated weight gain: management strategies and impact on treatment adherence', in *Neuropsychiatr Dis Treat* (2017).

You can read details on how beta blockers (medication frequently used to lower blood pressure or heart rate) can lead to weight gain in the article 'b-Adrenergic Receptor Blockers and Weight Gain, A Systematic Analysis', by Arya M. Sharma, Tobias Pischon, Sandra Hardt, Iris Kunz and Friedrich C. Luft, in *Hypertension* (2001).

The Dutch organization Wemos has created an informative online fact sheet (in Dutch) on hormone disrupters: https://www.wemos.nl/wp-content/uploads/2016/11/Wemos-Factsheet-Dit-moet-je-weten-over-hormoonverstorende-stoffen_November-2016-1.pdf.

The Endocrine Society, the American professional association for hormone specialists (endocrinologists), published an important article in 2015 on the effects of hormone disrupters and the risk of obesity and diabetes: 'The Endocrine Society's Second Scientific Statement on Endocrine-Disrupting Chemicals', in the leading journal *Endocrine Review*: https://endocrinenews.endocrine.org/edcs-linked-to-rising-diabetes-obesity-risk/. The Endocrine Society also provides tips on what you can do to prevent exposure to hormone disrupters as much as possible. You can find these online at https://www.endocrine.org/topics/edc/what-you-can-do.

On the website of The Endocrine Society you can find more information about how the effects of hormone disrupters can disrupt our appetite and affect our metabolism, and can potentially even lead to negative effects that can be passed on from generation to generation: https://www.endocrine.org/topics/edc/what-edcs-are/common-edcs/metabolic.

In 2017, Philippa D. Darbre wrote an interesting article entitled 'Endocrine Disruptors and Obesity' in *Current Obesity Reports* about the 'vicious spiral' that can occur if you have excess fat that can store hormone disrupters and so lead to a variety of diseases such as cancer.

A team of scientists led by Jeffrey Gordon from Washington University School of Medicine in St. Louis, Missouri published their ground-breaking study 'Gut Microbiota from Twins Discordant for Obesity Modulate Metabolism in Mice' in 2016 in the prestigious journal *Science*. The article describes experiments that demonstrate that traits such as obesity or leanness can be transmitted by transferring human gut bacteria to mice.

The following articles make for interesting scientific reading on the role of gut bacteria in obesity and stool transplants as one of the potential therapies: 'Gut microbiota and obesity: implications for fecal microbiota transplantation therapy' by Y. Kang and Y. Cai in *Hormones* (2017) and 'Transfer of intestinal microbiota from lean donors increases insulin sensitivity in individuals with metabolic syndrome' by A. Vrieze et al. in *Gastroenterology* (2012).

The following are other recent (2018) good review articles published about the role of our gut bacteria in obesity: 'The Gut Microbiome Profile in Obesity: A Systematic Review' by Olga Castaner et al. in *International Journal of Endocrinology* and 'Evaluating Causality of Gut Microbiota in Obesity and Diabetes in Humans' by Meijnikman, A.S., Gerdes, V.E., Nieuwdorp, M. and Herrema, H, in *Endocrine Reviews*.

If you would like to read more about Dr Richard Atkinson's assertion that viruses might also be contributing to the global obesity epidemic, his article 'Obesity Due to a Virus: How this Changes the Game' is available online on the website of the Obesity Action Coalition: https://www.obesityaction.org/community/article-library/obesity-due-to-a-virus-how-this-changes-the-game/.

In addition, already in 2005 Richard Atkinson et al. published a scientific article about the association between the adenovirus 36 and obesity, in the article 'Human adenovirus-36 is associated with increased body weight and paradoxical reduction of serum lipids', in the *International Journal of Obesity*. In 2015, a review article that looked at studies in recent years on this virus and its relationship to obesity, entitled 'Human Adenovirus 36 Infection Increased the Risk of Obesity: A Meta-Analysis Update' by M.Y. Xu et al., was published in *Medicine*.

Research on the use of antibiotics in early childhood and the negative effects on the microbiome and the association with higher body weight is described in the article 'Intestinal microbiome is related to lifetime antibiotic use in Finnish pre-school children' by Katri Korpela et al. in the leading journal *Nature Communications* (2016).

10 HOW CAN WE TACKLE OVERWEIGHT EFFECTIVELY?

If you would like to read a more detailed account of the study on the effect of daily intake of a sugar-sweetened beverage on the weight of schoolchildren, you can find this in the following article: 'A trial of sugar-free or sugar-sweetened beverages and body weight in children' by J.C. de Ruyter et al. in the *New England Journal of Medicine* (2012).

European practical and patient-centred guidelines for adult obesity management in primary care have been published by Dominique Durrer Schutz et al in *Obesity Facts*, 2019; 12:40–66. Recommendations on which treatment should be used for which degree of overweight or obesity is also contained in the Dutch health care standard for obesity (Zorgstandaard Obesitas), which was drawn up by the 'Partnership Overweight Netherlands' (Partnerschap Overgewicht Nederland, or PON) in 2010. The PON is the umbrella organization of professional associations of medical doctors and paramedical professionals, the Dutch association of health insurers Zorgverzekeraars Nederland (ZN), GGD-GHOR (the Association of GGDs (Community Health Services) in the Netherlands) and patient associations, which are all involved in obesity care in the Netherlands and also have an advisory role to the Dutch Ministry of Health, Welfare and Sport. You can find these Dutch-language resources online using the search terms 'Zorgstandaard Obesitas' and 'Partnerschap Overgewicht Nederland' or through this website: http://www.partnerschapovergewicht.nl/images/Organisatie/pon_Zorgstandaard_Obesitas_2011_A4_v1_04.pdf.

How to diagnose the underlying factors that contribute to becoming or remaining overweight is described in the following articles: 'Obesitas in de spreekkamer. Eerst diagnostiek en daarna effectieve behandeling (klinische les)' ['Obesity in the consultation room. First a diagnosis, then effective treatment (clinical lesson)'] by Eline S. van der Valk, Mesut Savas, Jan Steven Burgerhart, Maaike de Vries, Erica L.T. van den Akker and E.F.C. van Rossum in *het Nederlands Tijdschrift voor Geneeskunde* [*Dutch Journal of Medicine (NTvG)*] (2017) (161:D2310) as well as in the international article 'A comprehensive diagnostic approach to detect underlying causes of obesity in adults', by Eline S. van der Valk, Erica L.T. van den Akker, Mesut Savas, Lotte Kleinendorst, Jenny A. Visser, Mieke M. van Haelst, Arya M. Sharma and Elisabeth F.C. van Rossum in *Obesity Reviews* (2019). These are scientific articles. For more accessible information for the general public, you can also watch Liesbeth van Rossum's inaugural lecture (in Dutch), entitled 'Dik ben je niet voor de lol' ['You're not fat for the fun of it'], or her TEDx presentation 'Solutions for the obesity epidemic', both available online.

If you would like to read about the research on vaccination against the hunger hormone ghrelin that slowed weight gain in rats, more information is contained in the article 'Vaccination against weight gain' in *Proceedings of the National Academy of Sciences* (2006), by Zorrilla, E.P. et al.

Would you like to check whether you are a heavy or excessive drinker? You can find this information at the website of Centers for Disease Control and Prevention (CDC): https://www.cdc.gov/alcohol/fact-sheets/alcohol-use.htm.

11 FAT-SHAMING AND THE PSYCHOLOGICAL CONSEQUENCES OF OBESITY

Eet mij [*Eat Me*] is an interesting book (in Dutch) about the psychology behind eating and overweight, written by Asha ten Broeke and Ronald Veldhuizen. Asha ten Broeke was mentioned in Chapter 11 and is a psychologist and columnist for the Dutch daily newspaper *de Volkskrant*. Ronald Veldhuizen is a biologist and science journalist.

An extensive review of decades of research into prejudice against and stigmatization of people with overweight and obesity is contained in Puhl, R.M. and Brownell, K.D., 'Bias, discrimination, and obesity'. *Obesity Research* (2001), 9:788-905.

You can read more about how we deal with the obesity stigma in the article 'Weight bias: a call to action' in *Journal of Eating Disorders* (2016), by Angela S. Alberga, Shelly Russell-Mayhew, Kristin M. von Ranson and Lindsay McLaren.

More details about the findings that obesity can lower the likelihood of being hired for a job are contained in the scientific article 'Obesity Discrimination in the Recruitment Process: You're Not Hired!' in *Frontiers in Psychology* (number 7, 2016), by Stuart W. Flint, Martin Čadek, Sonia C. Codreanu, Vanja Ivić, Colene Zomer and Amalia Gomoiu.

An extensive review of the biological explanations behind the association between obesity and depression can be found in our article 'Depression and obesity: evidence of shared biological mechanisms', by Yuri Milaneschi, Kyle Simmons, Elisabeth (Liesbeth) F.C. van Rossum and Brenda Penninx, in *Molecular Psychiatry* (2018).

The following is a fascinating review article about how our gut communicates with the brain: 'The Role of Nutrition and the Gut-Brain Axis in Psychiatry: A Review of the Literature' in *Neuropsychobiology* (2018), by Sabrina Mörkl, Jolana Wagner-Skacel, Theresa Lahousen, Sonja Lackner, Sandra Johanna Holasek, Susanne Astrid Bengesser, Annamaria Painold, Anna Katharina Holl and Eva Reininghaus.

More information about the concept of a type of 'leak' of gut bacteria and other substances that can reach the brain through the bloodstream is contained in the review article 'Leaky Gut, Leaky Brain?' by Mark E.M. Obrenovich, in *Microorganisms* (2018).

If you would like to know more about how the symptoms of depression can improve by losing weight, we highly recommend this scientific article: 'Intentional weight loss and changes in symptoms of depression: a systematic review and meta-analysis' by A.N. Fabricatore et al., in *International Journal of Obesity* (2011).

Would you like to read more about the psychological consequences of bariatric surgery as an obesity treatment? This is described in the article 'Psychological Aspects of Bariatric Surgery as a Treatment for Obesity' by Sandra Jumbe et al., in *Current Obesity Reports* (2017), 6(1): 71-78.

More information on how bariatric surgery can lead to an increased risk of problematic alcohol use can be found in the review article 'Alcohol and Drug Use Among Postoperative Bariatric Patients: A Systematic Review of the Emerging Research and Its Implications' by Spadola et al., in *Alcoholism: Clinical and Experimental Research* (2015), 39(9):1582-1601.